Physiological Genetics of Agricultural Crops

Other Related Books

Plant Breeding : Theory & Practices 2nd Revised & Enlarged Edition	*Neal C. Stosckopf*
Genetic Improvement of Field Crops	*C.B. Singh*
Laboratory Manual of Microbiology Biochemistry and Molecular Biology	*Saxena, I.*
An Introduction to Population Genetics	*Crow, J.F.*
Biotechnology Emerging Trends	*Sayyed, R.Z.*
Plant Molecular Genetics	*Chakraborty, S.*
Physiology of Nutrition and Environmental Stress in Crop Productivity	*A. hemantaranjan*
Crop Improvement and Mutation Breeding	*A.K. Sharma*

Physiological Genetics of Agricultural Crops

Andor Balint

United Book Prints

Reprinted in India by:

UNITED BOOK PRINTS
(Imprint of Scientific Publishers, INDIA)
5A, New Pali Road, P.O. Box 91
Jodhpur 342 001, India
E-mail: info@scientificpub.com
Website: www.scientificpub.com

Print: 2021

1st Reprinted in India, 2015

ISBN: 978-93-90199-73-0

Printed in India

Contents

Preface

The study of the physiological genetics of field crops has been included in the curricula of many universities in recent years.

In the first part of my book "Production genetics of field crops", published in Hungarian in 1977, I summed up the major results of physiological genetics investigations, extending its scope in some respects. This English edition has now been enlarged and revised.

This branch of science covers the bases of physiological processes most important from the point of view of productivity, their relationship and—wherever possible—their genetic rules. Factors shaping the phenotype are examined using systems analysis. In Wallace's words, our aim is—"to modify the breeders' concepts of the ideal type of plant and the growers', agronomists' and horticulturists' views on how to maximize productivity".

In developing the ideal type of plant, a major task is to improve energy utilization. The very low percentage of solar energy utilized by the field crops in the course of photosynthesis is not the only point in question; Rabson et al. (1978) pointed out that a spectacular increase in the yield averages "was achieved by a high supplementary energy input in the form of fertilizer, fuel and traction power, and in many cases the returns were not in proportion with the expenditure of energy".

Physiological genetics is not the only basis for solving the tasks of plant breeding. It is of use in combination with molecular-, biochemical-, Mendelian- and statistical genetics for supplementing their data and achievements.

This branch of science has made considerable advances only in the past decade, explaining why not all questions raised can be answered satisfactorily. Nevertheless, it is hoped that the book will be helpful in developing a new approach and in increasing the number of those who will contribute to the progress of this field of science.

Dr. Andor Bálint

University of Agricultural Sciences
Gödöllő, Hungary

Acknowledgements

I wish to thank the following authors, institutions and publishers for their permission to reproduce the following material:

Masson, Paris (Figures 1, 2, 3), Academic Press, Inc. New York (Figures 6, 40), American Society of Agronomy, Madison, Wisconsin (Figures 9, 15, 18, 19, 20, 21, 25, 27, 28, 29, 30, 34), Cambridge University Press (Figures 23, 41, 42), American Seed Trade Association, Washington, (Figure 14); Publisher "Stinica", Kisinev (Figure 24), Iowa State College Press, Ames, Iowa (Figure 44), Paul Parey Verlagbuchhandlung, Berlin (West), (Figure 46), Brookhaven National Laboratory, Upton, New York (Figure 13).

CHAPTER 1

Physiological genetics and its relation to other fields of genetics

Physiological genetics is concerned with genetic determination of the life processes interacting with the environment and resulting in the full development of organisms (of field crops in our case). The same end is, in fact, pursued by the science of phenogenetics: to explain the successive stages of development at the level of genetic regulation.

A lot of works have appeared on the interaction between the different phases of ontogeny/environment. (Kuperman 1977). Not only the individual organs but the role of growth tip differentiation in the formation of various organs too have been studied without trying to find genetic correlations.

In addition to outlining the processes of cell and tissue differentiation Demarly (1977) has shown how the regulatory genetic system acts on the different part processes, according to Vincent (1970).

The informatory system responsible for the building-up of organisms acts through the division of cells and—as a result of tissue differentiation—the functional differentiation. Both processes take place in space and time. The cell is not able to divide and grow at the same time. Cell division is controlled by auxins and cytokinins, as well as by certain elements, e.g. calcium. During mitosis, the rate of ribonucleic acid synthesis is highly restricted. These changes are shown in Fig. 1.

Cells already differentiated may in some cases (e.g. after injuries) divide further.

Tissue development is based on the apical meristem. As a matter of fact, the zones of intensive cell division are included in the apex. As for the frequency of mitosis can be continuous, but certain cells of trees e.g. are able to remain in a state of rest even for 10 years before undergoing division, while in the active zone mitosis may occur daily in each cell.

Let us outline the development of a plant from an axis consisting of a single cell. In respect of the rate of division we distinguished stable zones:

1/100 for the last cell (a_1), 1/6 for the second-, 1/3 for the later one, and a rate of 1/1 for a third type, per unit time (it may be 1 day).

Observations made on this simple theoretical plant can be excellently applied to actual cases (Fig. 2)

According to Vincent (1970) this system functions under the control of the phytochrome-ascorbic acid system shown in Fig. 3.

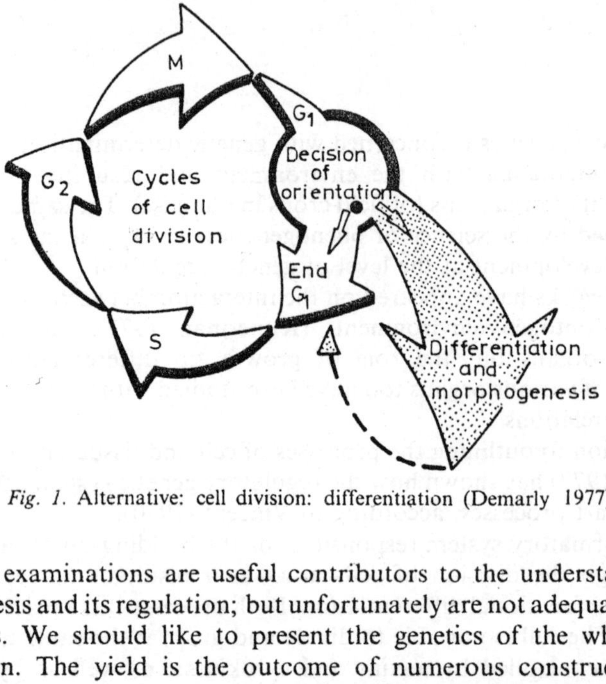

Fig. 1. Alternative: cell division: differentiation (Demarly 1977)

These examinations are useful contributors to the understanding of ontogenesis and its regulation; but unfortunately are not adequate for our purposes. We should like to present the genetics of the whole yield formation. The yield is the outcome of numerous constructing and decomposing processes in interaction with each other, the roles and interrelations of which constantly change during ontogenesis. The aim of the genetist is to attain a favourable harvest index, the highest possible proportion of useful yield (grain or tuber), with as low a material and energy input as possible.

The required method should supply information not only on the components of end products (plant, spike, etc.) but also on the life processes leading to their development and on the relations of these life processes.

This can be achieved by using systems analysis giving information on the trends of input (elements introduced) and output (performance, end

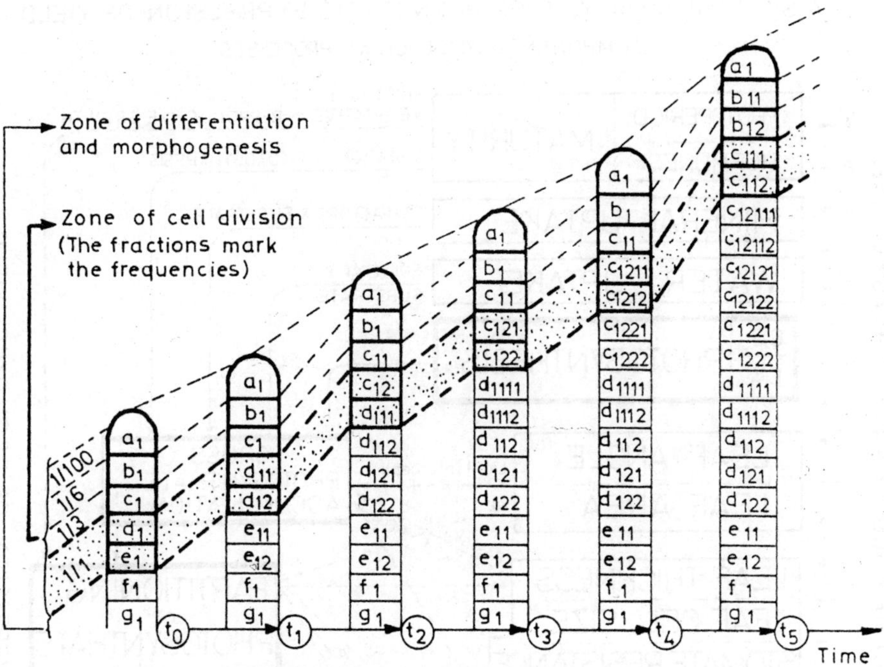

Fig. 2. Diagram of apical growth (Demarly 1977)

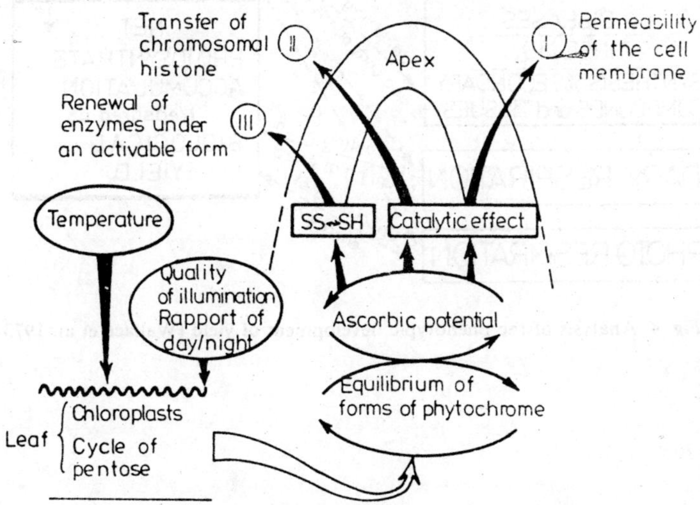

Fig. 3. Complex sketch of correlations (Vincent 1970)

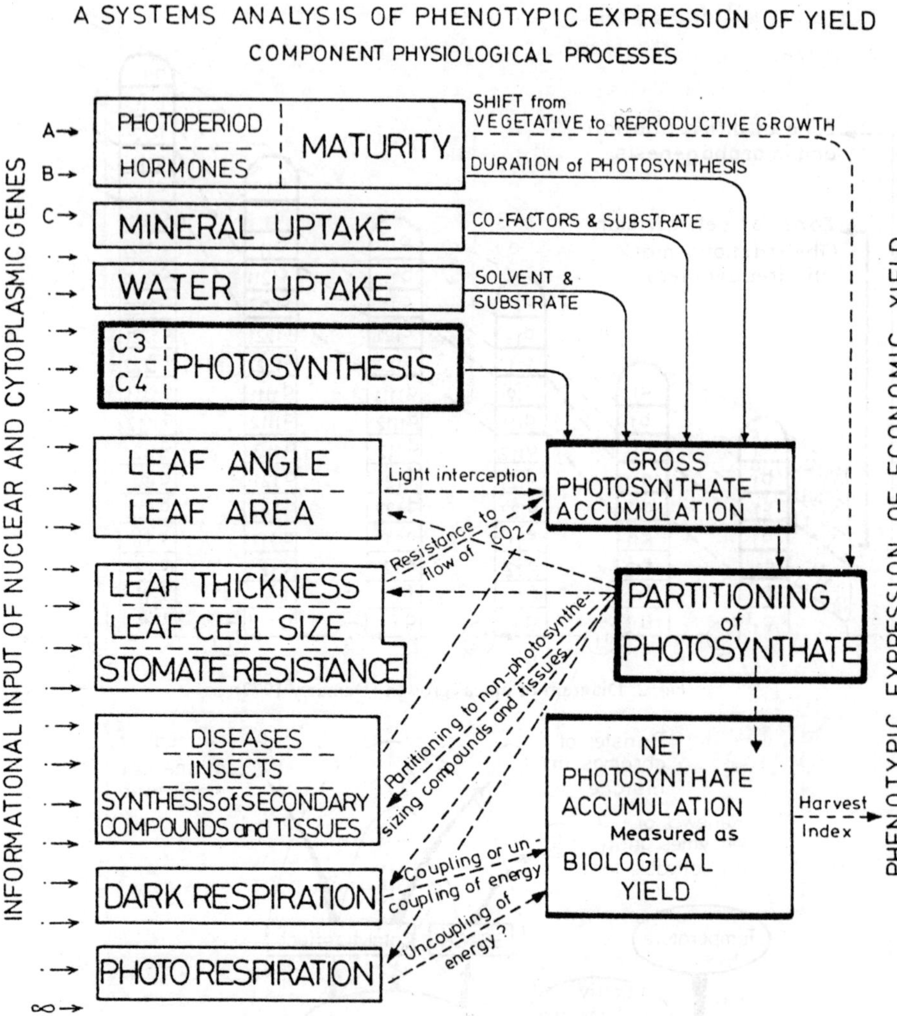

Fig. 4. Analysis of the phenotypic development of yield (Wallace et al. 1973)

product), as illustrated by Wallace et al. (1973) in Fig. 4. Large amounts of organic matter (biological output) have to be produced with the highest possible crop yield (harvest index). This requires an increased photosynthetic activity using a minimal amount of synthetized organic matter in the form of energy in the life processes and maximal incorporation of it in the yield.

By analysing these part processes in the way with respect to a plant species we may approach the ideal type of variety.

Scope and constraints of mathematical genetics

Mendelian genetics gives reliable information on the hereditary nature of qualitative features (green or yellow colour of pod, full or shrunken seed in pea etc.), and when crossing parents with such characteristics the breeder can safely reckon on the appearance of the desired neocombination in the F_2. These qualitative features are, at the same time, quantitative characters too, since e.g. the amount of starch per 1 kg dry matter is 350 g in peas with round seeds, and 200 g in those with shrunken seeds.

Unfortunately, it occurs very rarely that a quantitative feature is determined by a single gene (polygenic transmission is characteristic of a high percentage of the cases) thus the inheritance of quantitative characters is studied with statistical genetic methods (mean values and variations being the basis). Conclusions drawn from data of large populations can be utilized in many respects, for they enable genetic- and environmental variation to be separated supply information on certain types of interaction between the genes (additive, epistatic, etc. effects of genes).

This information refers, however, only to the joint phenotypic effect of all genes and not to the actions of the individual genes. It is of no avail for the breeder to know that the protein content in maize is determined by 22 genes (Frey 1949); gene coupling, plasmic effects cannot be taken into account although as for the latter e.g. the distribution of the T-race of *Helminthosporium maydis* has quite recently (Hooker 1967) called attention to the fact that susceptibility may be a plasmically determined characteristic.

In preparing selection programmes the h^2 value is good indicator of the expectable progress. In the F_2 populations more than 60 per cent of the variation is of genetic origin, providing a good chance of selecting the right genotypes.

Table 1

Heritability estimates in percent
(Brim 1973)

Character	Johnson and Bernard (1963)	Anand and Torrie (1963)			Kwon and Torrie (1964)		Fehr and Weber (1968)*		Smith and Weber (1968)		Byth et al. (1969)+	
		Cross			Cross		Cross		Cross		Cross	
		1++	2	3	4	5	6	7	7	8	7	8
Yield	38+++	23	33	50	10	3	39	28	52	57	58	50
Seed weight	68	53	65	84	44	79	92	94	92	93	88	91
Height	75	82	84	73	70	67	66	85	82	80	90	86
Lodging	54	59	43	61	51	74	60	75	63	68	70	68
Days to flowering	84	91	65	87	75	76	—	—	—	—	—	—
Fruiting	65	67	46	48	66	81	—	—	—	—	—	—
Maturity	78	84	81	86	79	82	75	75	90	91	92	94
Protein	63	—	—	—	57	—	76	77	86	90	81	88
Oil	67	—	—	—	51	—	74	72	88	89	82	86

* Based on one environment only.

\+ Averaged over maternal and daughter lines.

++ Same numeral on cross designation indicates progenies were from crosses of the same parentage.

+++ All estimates were obtained from variance components except for those of Johnson and Bernard who presented "expected heritabilities" based on available data and observation of several soybean breeders and geneticists.

Attention should be paid to the following in connection with the h^2 values. Statistical genetics usually works with estimated values, established for a given population in the case of the h^2 value, too, explaining why we find different values for the same characteristics (Brim 1973, Table 1).

According to the evidence of the data the value of h^2 is the lowest for the total yield, which agrees with the experiences of soybean breeders too. Differences are, naturally, encountered in certain combinations in this case too, since productivity is shaped by the joint action of almost every gene of the organism—in interaction with the environment. In crossings carried out with the variety Capelle-Deprez it was pointed out by Lupton (1966) that inheritance for productivity was dominant beside a high variability of yield components. This, however, applies at the most to the mentioned variety and the given crossing combinations.

Physiological genetics cannot be substituted for mathematical genetics it only explains and completes its findings from time to time, and widely utilizes its results in defining and interpreting its own regularities. A similar interrelation exists between biochemical and mathematical genetics.

Relations between molecular and physiological genetics

The first stage of genetic control is the input of the gentic information of nucleus and plasma. This entails the synthesis of special enzymes which control the individual life processes in the presence of appropriate ions, co-factors and environmental conditions. Each enzyme catalyzes a special step of synthesis or degradation. The enzymes themselves may be degraded and small enzyme molecules synthetized into larger ones.

The plants, like other living organisms, have the peculiarity that the same compound can catalyze the synthesis of two or more protein molecules of un-identical composition. They are charged with electricity, and can be separated by electrophoresis according to molecular weight, size or shape. Changes usually occur in the amino acid sequence and do not affect their catalytic activity. They are therefore called isoenzymes.

They are biologically important because their effectiveness varies both quantitatively and from the point of view of reaction to the environment. In response to environmental factors—to temperature in particular—the activity of almost every enzyme changes (according to the well-known rule of Q_{10} a 10 °C rise in temperature redoubles the activity of enzymes), whereby the activity of practically every gene changes. In the cou' e of the physiological processes this develops a new interaction genes, influences the formation of substrates catalyzed by the enzymes and thus causes the repression or depression of one or more genes. The isoenzymes do not give uniform responses to these changes, varying with different concentrations or temperature. Schwartz (1962) found in his experiments that hybrid forms of the esterase enzyme could be identified in maize hybrid while the isoenzymes of both parents were dominantly transferred to the progeny. In other cases intermediary inheritance was observed.

There are differences in the activity of isoenzymes between the species during ontogenesis, and between the tissues (Hess 1968). Different isoenzymes take part in the processes of root, shoot and flower formation, and their activity changes in the course of endosperm or pollen formation too. Through the functions of genes, alleles, suballeles they control the development of the organisms with their parallel and successive—"series-connected"—activity.

Table 2

Isocitrate-dehydrogenase activity
(mg/k TTH/100 seedlings) (Sysoev and Semenyuk 1975)

Variety	1968	1969	1970
Koperatorka	243.0	264.2	261.2
Odesskaya 3	300.5	322.3	363.7
Novostepnaskaya	320.5	335.0	316.3
Yuzhnoukrainka	315.1	325.6	325.8
Bezostaya 1	376.2	408.8	393.8
Odesskaya 51	386.3	408.5	403.2
Priboi	395.1	418.4	449.2
Mironovskaya 808	384.3	394.9	421.6
Avrora	395.3	386.6	395.3
$SD_{0.95}$	58.2	57.4	64.1

Phenogenetic research will no doubt contribute to solving the tasks of physiological genetics. At present, however, the phenogenetics of higher plants is just beginning to take shape, and can explain the part processes of genetic determination during the ontogenesis rather than the whole development (Hess 1974).

For lack of sufficient studies mostly only the variations between and within the species will be pointed out. Attempts have been made to draw conclusions from the data of enzyme activity studies on the productivity of improved varieties and strains. Sysoev and Semenyuk (1975) start from the principle that the rate of biosynthesis is the indicator of the intensity of metabolic processes, and at the same time, of higher productivity. Having studied a number of enzymes they found isocytrate-dehydrogenase to be the most suitable for demonstrating this correlation. The enzyme activity was determined in 4-day-old etiolated seedlings by a method using tetrazolium. In varieties with increasing productivity (Avrora was the most productive) the enzyme analyses gave the following results (Table 2, Sysoev and Semenyuk 1975).

In this experiment the authors found a significant correlation of r= +0.676 in 1970 and r= +0.614 in 1971 between productivity and enzyme activity. From this they tried to make forecasts on the productivity of strains within a variety. In 1970 and 1971 they studied the responses of 32 and 41 biotypes, respectively, in the variety Odesskaya 51. The value of r was +0.506 in 1970 and +0.488 in 1971. This correlation is not adequate for making a reliable preselection. The authors themselves drew the conclusion that other investigations (e.g. intensity of photosynthesis) would be needed to establish reliable correlations.

CHAPTER 2

Genetics of physiological yield components and agronomical characters

Genetics of photosynthesis

Attainment of higher and better quality yields is based on the photosynthetic ability of plants (Pál 1973).

Under field conditions the intensity of photosynthesis depends on the mineral nutrient and water supply of plants, the activity of pathogens and pests, the age of the plant and leaves etc.—in addition to illumination, temperature and CO_2 content of the air. The global energy utilization of photosynthetically active radiation (PAR) for biomass production is very low: 0.15 per cent. PAR comes from the spectral range of 380–720 nm. The photosynthetic pigments utilize different parts of it. In the temperate zone on a clear summer day solar radiation supplies 500 cal/cm^2 (the surface of earth is reached by 2.5×10^{20} cal a year), of which the share of PAR is 222 cal/cm^2. The amount of red light required to fix 1 molecule of CO_2 is 440 kcal (11 Einsteins). Results of caloric measuring show that in the reduced CO_2 molecule 118 kcal chemical energy is bound. The theoretically possible utilization is thus $118/440 = 27$ per cent.

However, not all solar radiation is photosynthetically active. For crop plants the range is about 400–690 nm. This represents less than half the total energy flux, which is basically the visible spectrum (44%). In addition, a 10% reflectance loss must be accounted for. Plant respiration accounts for a further loss of about 10% which may be a conservative estimate. The calculation continues:

$$0.27 \times 0.44 \times 0.9 \times 0.9 = 10\% \text{ efficiency}$$

Taking into account all the factors the attainable efficiency is a mere 10 per cent. According to the calculations of Zelitch (1971) the maximum efficiency is 11–12 per cent. With maize 4.6 per cent has been found to be the maximum in Holland, 6.4 in California and 9.8 per cent in the USA.

In practice only a fraction of this possibility is realized. With adequate radiation the temperature may be low, the water supply inadequate, etc.

It should be noted that low efficiency of photosynthesis is not the reason preventing the solution of the world nutrition problem. This is clearly seen from the data of Wittwer (1974) who compared the average and record yields in the United States (Table 3).

The genetic possibility of multiplying the yield averages is given if the right conditions are provided for the plants.

Table 3

Average, high and record yields in the United States (kg/ha) (Wittwer 1974)

Plant	Average	High	Record	Record/ average
Maize	6016	14720	19584	3.25
Wheat	2048	8640	13824	6.75
Soya	1792	5120	7040	3.93
Sorghum	4032	12800	20480	5.08
Rice (Philippines)	1792	8320	10670	5.90
Potato	24640	64000	89600	3.64
Batata	11520	38400	57600	5.00
Barley	2624	9600	13568	5.17
Oat	3136	9600	18944	6.04
Sugar beet (ton/ha)	49.4	98.8	133.38	2.70

This is particularly the case in tropical regions where e.g. annually rice crops of 16.4 ton/ha have commonly been harvested, and the record has been as high as 26.0 ton/ha; the record yields have been: 77.0 ton of cassava. 41.0 ton of sweet potato from a single harvest. Of sweet potato as much as 65.2 tonnes have been produced a year (Alvim and Kozlowski 1977).

Differences of light utilization in species and varieties

The same tendency was established by Good and Bell (1978 in: Rabson et al. 1978) who gave the world averages and the results of the most developed countries for the percentage of energy incorporated in the different plants (Rabson et al. 1978, Table 4). The data reveal that the differences in solar energy utilization may be as high as 100–200 per cent. These differences are due partly to ecological—but mainly—to agrotechnical factors.

It is not indifferent how these values change in the course of the ontogenesis as a function of ecological conditions.

Changes in photosynthetic productivity during the successive development phases in two sunflower varieties can be seen in Fig. 5.

The figure illustrates that there are differences not only between the varieties but also between the crop years. In both years the Mayak variety is less sensitive to the environmental factors. In spite of this, the Peredovik variety is superior with respect to total organic matter production.

Apart from these effects solar energy utilization varies considerably within the species as well (Table 5, Zelitch 1971).

Since the products of various species are needed a species can hardly be replaced by another one. Those differences which appear among the varieties of a species (Table 5) are of much more significance. This is

Table 4

Efficiency of conservation of light energy in various crops
(Good and Bell, in: Rabson et al. 1978)

Crop	Energy conserved annually in products of photosynthesis (kcal/m^2 × 10^3)	Efficiency of use of sunlight* (%)
Wheat (The Netherlands)	4.4	0.35
Wheat (world average)	1.3	0.10
Corn (USA)	4.5	0.35
Corn (world average)	2.4	0.17
Rice (Japan)	5.5	0.42
Rice (world average)	2.3	0.18
Potato (USA)	4.1	0.31
Potato (world average)	2.2	0.17
Soybean (Canada)	2.4	0.18
Soybean (world average)	1.4	0.10
Sugar cane (Hawaii)	12.2	0.95
Sugar cane (Cuba)	3.7	0.30
Sugar beet (The Netherlands)	7.3	0.56

* Energy conserved each year in the biomass of the entire plant divided by the energy content of the sunlight falling each year on the crop lands (assumed to be about 1.3 × 10^6 kcal/m^2).

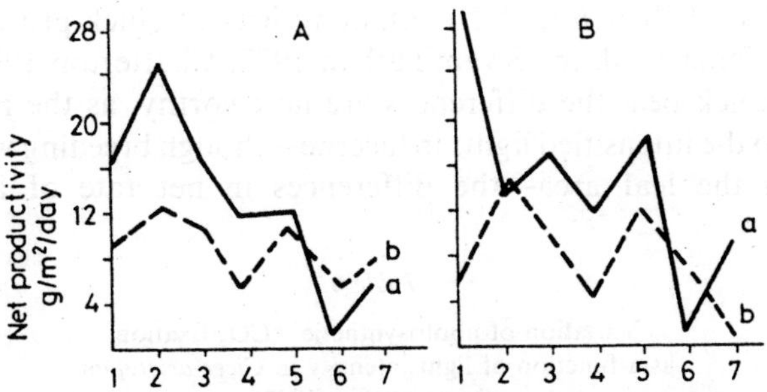

Fig. 5. Changes in the net productivity of photosynthesis in the course of ontogenesis in the sunflower (Baranovskii and Savka 1974). A = variety Peredovik; B = variety Mayak; a = 1968; b = 1969; 1–leaf formation, 2–differentiation, 3–active growth, 4–flowering of heads, 5–seed formation, 6–filling of seeds, 7–full maturity

Table 5

Typical rates and variations within species of net photosynthesis in single leaves of various species at high illuminescence and 300 ppm of CO_2 in air (Zelitch 1971)

Species	Net photosynthesis ($mg\ CO_2/dm^2/h$)
C_4 plants	
Maize (15 inbreds and hybrids)	28–85
Maize (20 races)	48–59
Sugar cane	42–49
Sorghum	55
Pigweed (*Amaranthus edulis*)	58
C_3 plants	
Sunflower	37–44
Cattail (*Typha latifolia*)	44–69
Tobacco (3 varieties)	11–17
Sugar beet	24–28
Rice (50 varieties)	34.5–62
Barley (7 varieties)	17.7–21
Soybean (36 varieties)	12–24
Soybean (20 varieties)	29–43
Wheat	17–31
Bean (5 varieties)	13.7–18.5
Gossypium (26 varieties)	24–40
Oak	10
Maple	6
Camissonia	94

illustrated in Tables 6 and 7 containing lists of chick-pea and lucerne varieties (Shina et al. in: Swaminathan 1972, Chatterton 1976).

In the chick-pea, the differences are noteworthy, as the responses of varieties to the intensified light; in lucerne—though breeding was aimed at increasing the leaf area—the differences in net rate of assimilation

Table 6

Variation of photosynthetic $^{14}CO_2$ fixation as a function of light intensity in *Cicer arietinum* (Swaminathan 1972)

Variety	$^{14}CO_2$ fixation (10 min/g fresh weight/1 g)	
	900 foot–candle	9,000 foot–candle
BSS 1	13,000	17,200
G 62-404	17,400	22,900
G 130	12,200	15,900
H-208	14,400	15,200
H-214	15,100	12,700
H-534	14,400	18,700
JG-101	20,700	31,100
L-144	14,700	20,000
L-345	12,000	34,600
N-59	20,700	18,800

(measured by CER) are smaller, significant difference is only found in the leaf area of a few varieties.

The distinction between the C_3 and C_4 plants in Table 5 must be briefly explained. In the secondary processes of photosynthesis the C_3 plants are characterized by the Calvin-cycle, the C_4 plants by the dicarbonic acid reaction pathway of CO_2 fixation, as pointed out by Hatch and Slack (1966 in: Hess 1979) first in maize and sugar-cane, then in other mono- and dicotyledonous plant species too. This could result in a net photosynthesis higher than in the C_3 plants, and from an ecological viewpoint enables the plants to carry on efficient photosynthesis, even with closed stomates, in a water deficient environment by using CO_2 formed in the course of photorespiration (Hess 1979). A further advantage is that the C_4 plants are able to utilize light more intensively than the C_3 type (Fig. 6, Ludlow and Wilson 1971, in: Alvim and Kozlowski 1977). The question arises what the maximum potential value is, and how it can be attained. Many countries are conducting experiments and trial production with alga cultures, and produce 8–10 g/m² dry matter a day (Tamy 1957 in: Nichiporovich 1967). The product thus obtained, though very rich in protein, has not the advantage of quantity over crops grown in the field. The results of experiments carried out by Moskov (1963 in: Nichiporovich 1967) show that under favourable conditions the tomato can exceed the output of algae three- to fourfold. In his experiments Moskov used an artificial medium (keramzit) instead of soil, so he could adjust the composition of the culture fluid to the demands of plants. He used artificial illumination for 14 hours a day (in a glasshouse light chamber) and placed 36 plants of the determinate Pushkinskii 1853 variety on an area of 1 m². Under optimum temperature and air humidity conditions provided for the tomato plants an average yield of about 500 g/plant—18

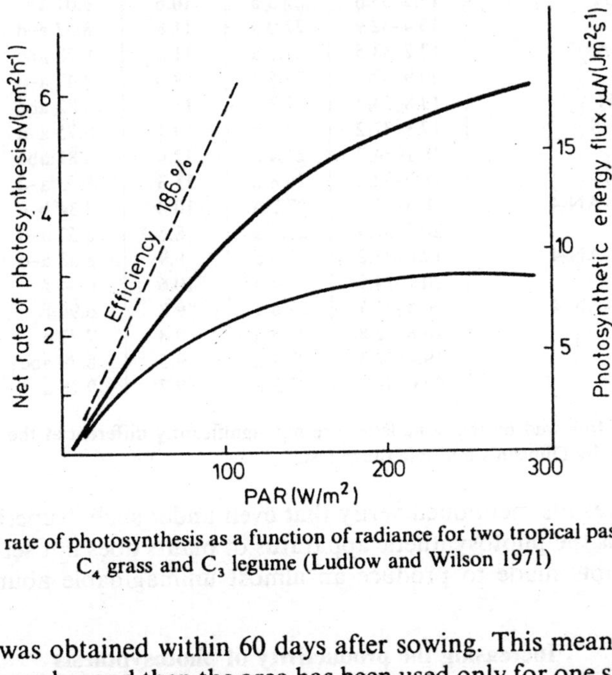

Fig. 6. Net rate of photosynthesis as a function of radiance for two tropical pasture species, C_4 grass and C_3 legume (Ludlow and Wilson 1971)

kg/m²—was obtained within 60 days after sowing. This means some 18 waggons per ha, and then the area has been used only for one sixth of the year.

According to Moskov this is far from the attainable possibilities, for with the method of individual selection made in artificial cultures a stock material capable of yielding 600–700 g per plant in 60 days has been produced.

Table 7

Carbon dioxide exchange rates (CER), leaf area (LA), and coefficients of variation (CV) measured on 22 alfalfa populations (Chatterton 1976)

Entry	CER, (mg $CO_2/dm^2/h$)			LA, (cm^2/leaf)	
	Range	Mean	CV %	Mean	CV %
BIC-5	20.3–32.7	27.3 a	8.5	8.41 bcd	18.8
BIC-5, AN	20.9–31.8	26.9 a	9.7	9.00 abc	24.5
BIC-5, BW	18.3–34.3	27.3 a	11.8	8.04 de	22.3
BIC-5, CLS	18.3–33.2	27.4 a	9.1	8.69 a–d	26.3
BIC-5, FLS	17.4–33.0	27.4 a	10.6	8.52 a–d	20.7
BIC-5, PRR	17.3–32.8	27.1 a	11.9	8.30 cde	23.5
BIC-5, PLH	19.3–33.6	27.3 a	10.6	9,07 ab	22.1
BIC-5, PA	15.4–32.9	27.0 a	11.8	8.57 a–d	18.8
BIC-5, SAA	17.2–33.5	27.1 a	11.2	8.78 abc	26.7
BIC-5, SN	20.9–32,8	27.5 a	9.0	8.47 a–d	22.5
BIC–5, WH	19.8–34.7	27.3 a	10.7	8.82 abc	21.7
BIC-5, WHS	17.8–32.2	27.1 a	9.1	8.75 a–d	24.0
BIC-6	16.1–34.7	27.4 a	12.4	8.81 abc	24.4
Glacier	18.0–32.2	26.8 a	11.3	8.53 a–d	21.2
Beltsville 1-AN-4	18.3–31.5	27.2 a	10.0	8.34 b–e	23.8
Saranac	20.5–32.9	27.9 a	8.5	8.57 a–d	22.0
Beltsville 2-AN-4	19.6–33.2	27.5 a	9.9	8.83 a–d	25.4
Vernal	20.1–34.9	27.4 a	10.6	6.44 f	22.6
Beltsville 3-AN-4	20.5–33.7	27.6 a	9.8	6.96 f	20.8
WD-S-5	16.8–32.8	27.5 a	9.8	7.72 e	21.8
Team	19.4–32.3	27.4 a	9.5	8.79 abc	25.9
ARC	22.0–33.3	27.3 a	9.7	9.18 a	19.9

* Means followed by the same letter are not significantly different at the 0.05 level of significance by Duncan's multiple range test.

These results mentioned verify that even under such "superintensive" conditions the photosynthetic apparatus of plants does not set a limit to the attempts made to produce an almost unimaginable abundance of foods.

Increasing the productivity of photosynthesis

The productivity of photosynthesis can be increased in various ways:

1. Intensifying photosynthesis by agrotechniques (fertilization, irrigation, etc.). This has been discussed in Chapters 3 and 4.
2. Increasing the assimilating leaf surface (LAI) per unit area (L = leaf, A = area, I = index; unit area = 1 m^2 growth space);

3. Increasing the duration of the leaf area (LAD).

LAI can be raised in the simplest way by increasing the stand density. Apart from other factors, 100–200 per cent higher yield averages of wheat and maize would have been unimaginable without an increase in the stand density. The low plant number per unit area is usually one of the reasons for the small yields occurring in the agricultural practice (Harper 1977).

The effect of increasing the stand density on LAI is shown in the data of Gotlin and Pucaric (1973) (Table 8).

Table 8

Effect of plant density on leaf area index (LAI) of inbred lines from various maturity groups (Gotlin and Pucaric 1973)

Inbred line	Leaf area index (LAI)			
	Plant density (plants per ha)			
	66,000	80,000	100,000	133,000
L 282	1.06	1.34	1.70	2.06
L 255	1.91	2.28	2.60	3.61
W 153 R	1.66	2.16	2.41	3.28
10 US	2.62	3.12	3.63	4.46
8 US	2.57	3.21	4.07	5.18

There are differences in the lines both in the LAI and in their response to the extent of stand density. The yield increase (Table 9) is not, however, proportionate to the increase of LAI, moreover, there are cases—e.g. the line 8 US—when it is the highest with a plant number of 80,000 (73,600 on harvesting).

A twofold increase in the LAI resulted in a maximum 20 per cent growth of total yield, and there were treatments where yield increase did not even occur. Bunting's (1973) experiments clearly show that higher stand density (and higher LAI at the same time) increases the yield only to an optimum point (Fig. 7). Of the yield components it is the grain yield/ear that first becomes lower in response to increased stand density.

For the early hybrids examined the optimum plant number ranges from 6 to 12 per m^2.

The reduction in the proportion of the seed production can be explained with different causes. The netto assimilation rate decreases. At dense population the shading effect of the plants decreases the light supply of the

Table 9

Effect of plant density on grain yield of inbred lines from various maturity groups (Gotlin and Pucaric 1973)

Inbred line	Plant density (plants/ha)	Plant density in harvest time (plants/ha)	Yield of air dry grain (kg/ha)
L 282	66,000	57,900	2,644
	80,000	70,400	2,822
	100,000	81,000	3,222
	133,000	109,000	3,644
L 255	66,000	59,900	3,444
	80,000	69,600	3,733
	100,000	83,000	3,711
	133,000	95,700	4,067
W 153 R	66,000	58,600	3,022
	80,000	61,600	2,644
	100,000	75,000	2,756
	133,000	111,700	3,644
10 US	66,000	61,300	2,178
	80,000	73,600	2,244
	100,000	87,000	2,333
	133,000	105,000	2,622
8 US	66,000	58,600	2,622
	80,000	73,600	2,889
	100,000	79,000	2,578
	133,000	109,000	2,867

LSD 0.05%......5.21
0.01%......6.91

lower leaves and that way the photosynthesis productivity, too. The light, the nutrition and the water supply of the plants are more unfavourable at higher plant density than lower. The higher population density — as can be seen in Fig. 7 — is 80–120 000 plants/ha, it is much more than the maximum population density is applied by certain countries (Fig. 8).

By the higher stand density—occasionally 80–120,000 plants—as seen in Fig. 7 we mean a much larger number of plants per unit area than the maximum used in certain countries.

As a consequence of increased stand density besides the reduction of ear weight the proportion of sterile plants becomes considerably higher in

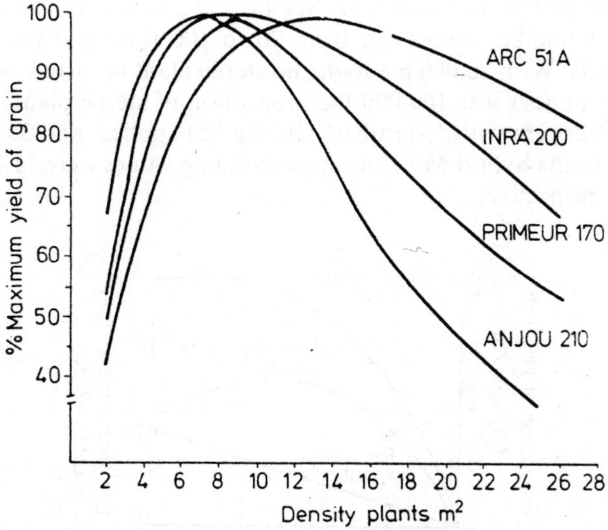

Fig. 7. Differential varietal response in grain yield to changes in density (Bunting 1973)

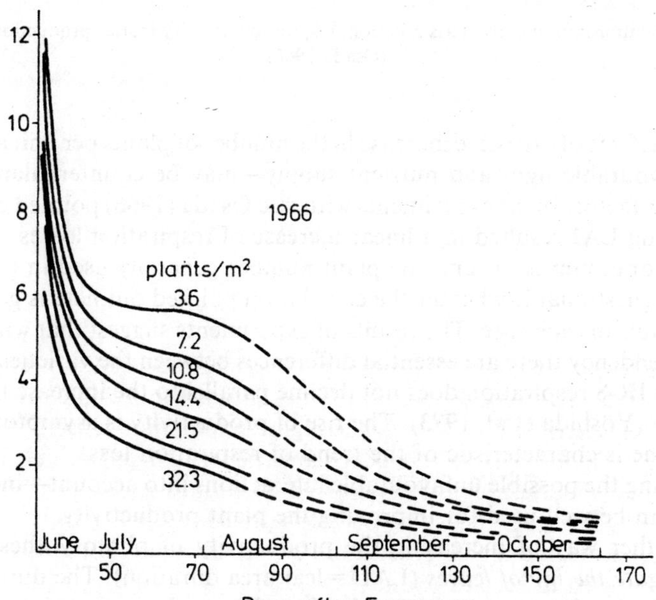

Fig. 8. The effects of density on seasonal trends in NAR (Voldeng 1971)

maize. I'só (1968) used an earlier (Mv 40) and a later (Mv 1) hybrid sown to a plant number increasing from 20 to 100 thousand per ha in his experiments. With 20,000 plants/ha no sterile plant occurred, while when the plant number was 100,000 the proportion of sterile plants of Mv 40 was 22.0 per cent with N-fertilizer (100 kg/ha) applied, and 35.4 per cent without. In the hybrid Mv 1 the corresponding values were 11.2 and 23.5 per cent, respectively.

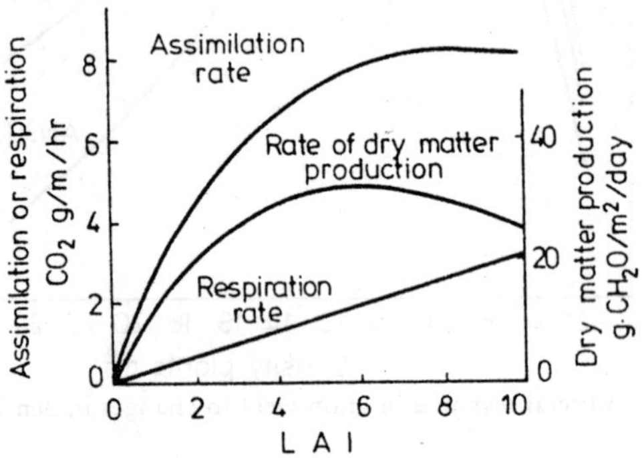

Fig. 9. Relationship of carbon assimilation, respiration and dry matter production to LAI (Osada 1966)

The effects of excessive increase in the number of plants per unit area—a less favourable light and nutrient supply—may be counterbalanced by another factor. In his experiments with rice Osada (1966) pointed out that the rising LAI resulted in a linear increase of respiration losses (Fig. 9).

The optimum is around the plant number generally used in practice.

It is questionable whether the correlation pointed out here is generally applicable in each case. The results of experiments suggest that within the basic tendency there are essential differences between the varieties. In the variety IR-8 respiration does not decline parallel to the increase in stand density (Yoshida et al. 1973). The rise of productivity is asymptotic, and the same is characteristic of the trend of respiration loss.

Taking the possible unfavourable interactions into account—increased LAI can be well used for improving the plant productivity.

Another way of increasing the productivity of photosynthesis is to *prolongate the life of leaves* (LAD = leaf area duration). The duration of leaves is a varietal feature just as the size of the leaf area that characterizes the lines and varieties. Borojevic (1973) showed the trends of LAI and

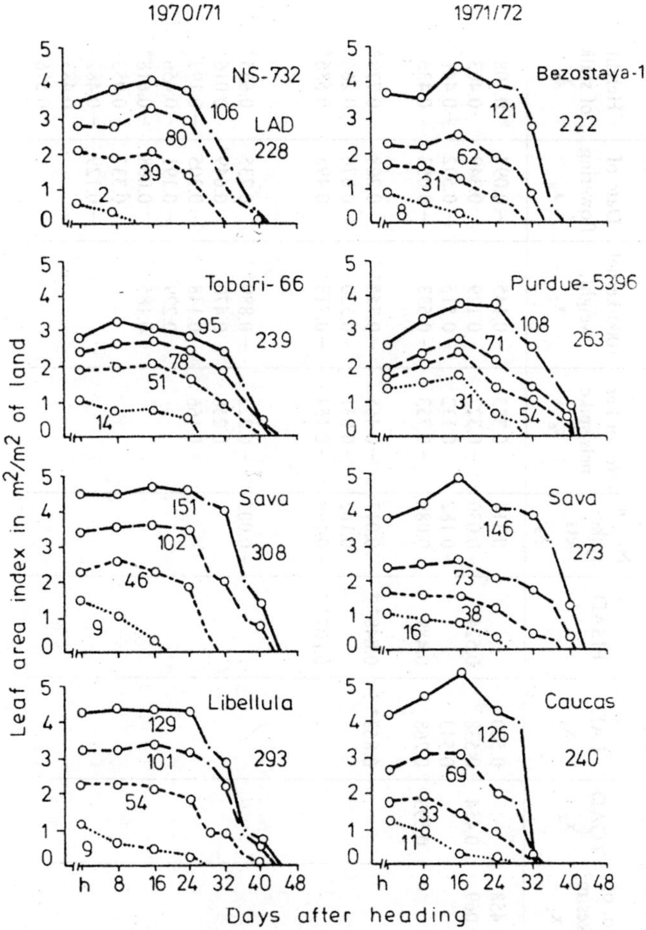

Fig. 10. Leaf area index of various leaves; – flag leaf, – – second, – – – third and . . . fourth leaf. Numbers represent leaf area duration in days (Borojevic 1973)

LAD for the varieties used in his experiments, and their leaves, one by one (Fig. 10).

Besides he studied the relationship between the plant structure and the yield components in 10 wheat varieties sown to a stand density of 500, 600 and 700 plants/m², respectively. The varieties Sava and Purdue developing a leaf area index of 10–12 were not found to be superior in productivity to

Table 10

Simple coefficient of correlation for ten characters in ten different genotypes (Borojevic 1973)

		No. of spikes/m² x_1	TGAD x_2	LAD x_3	FISAD x_4	No. of spike-lets x_5	No. of ker-nels/spike x_6	1000-kernel weight, x_7	Date of flowering x_8	Height of stalk x_9
Yield of grains	1	0.458		0.500		0.314	0.385	−0.145	−0.097	0.108
in mg/m²	2	0.069	0.454	0.538	0.524	−0.030	−0.327	−0.129	0.349	0.463
1 Number of spikes/m²	1			0.543		−0.182	0.132	−0.515	−0.592	−0.441
	2		0.593	0.268	0.489	0.089	−0.753	−0.533	−0.053	0.436
2 Total green area duration (TGAD)	2			0.863^{++}	0.848^{++}	0.017	−0.465	-0.755^{++}	0.208	0.779^{++}
3 Leaf area duration	1					0.162	−0.031	−0.530	0.076	0.266
(LAD)	2				0.807^{++}	−0.199	−0.154	-0.718^{++}	0.495	0.686^{++}
4 Flag leaf, top internode and spike area duration (FISAD)	2					−0.005	−0.323	-0.889^{++}	0.235	0.635^{+}
5 No. of spikelets	1						0.258	0.470	0.649^{+}	0.016
	2						0.166	0.148	−0.305	−0.199
6 No. of kernels per	1							0.229	−0.191	−0.456
spike	2							0.185	0.068	-0.618^{+}
7 1000-kernel weight	1								0.337	−0.055
	2								−0.129	−0.482
8 Date of flowering	1									0.483
	2									0.576

1 Data for 1970/71.
2 Data for 1971/72.
$^{+}$ and $^{++}$ significant for 5% and 1%.

Table 11

Development of leaf area in wheat
(Nichiporovich 1957)

Treatment	LAD	Total yield (kg/ha)	Grain yield (kg/ha)
1 Irrigation + fertilization	1.200	11,000	3,300
2 Irrigation	0.735	7,200	2,500
3 Control	0.345	3,400	1,100

LAD = leaf area duration.

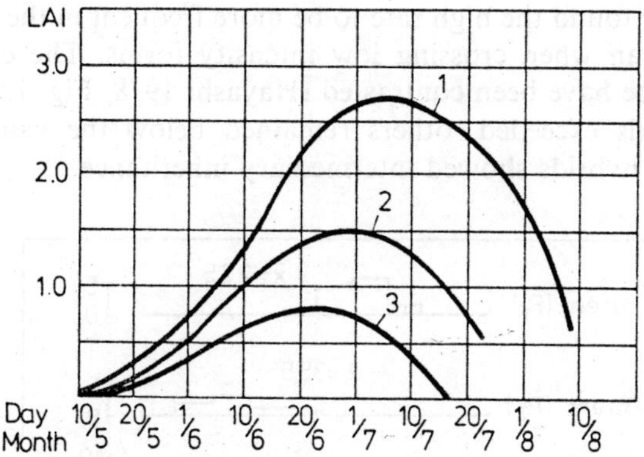

Fig. 11. Effects of irrigation-fertilization (1), and irrigation (2) on the leaf area index (LAI) and leaf area duration (LAD), control (3) (Nichiporovich and Stroganova 1957)

Caucas (Kavkaz), Heine and Bánkuti 1205—varieties with an LAI value of 8–10. A closer relation could be observed between the value of LAD and the volume of yield (Borojevic 1973, Table 10).

At the same time, there is a negative correlation between the value of LAD and the thousand-grain-weight. Genotypes with smaller leaves develop a larger number of spikes, but with a lower thousand-grain-weight. In genotypes with larger leaves it is the other way round.

Fertilization and irrigation equally increase the values of LAI and LAD. This relationship as found in experiments with wheat by Nichiporovich (1957) is shown in Table 11 and Fig. 11.

In experiments with wheat, substantial yield increases have been observed compared to the control. The larger yield can be explained—

beside the larger leaf area—by the longer life of leaves. Irrigation water without application of fertilizers while increasing the total volume of yield decreased the ratio of grain to straw.

The varieties naturally give diverse responses to the different treatments, as supported by data in Chapters 3 ıd 4.

Genetic analyses

The first task of genetic analyses is to determine inheritance in the *intensity level of photosynthesis*.

Loomis et al. (1971) crossing Mimulus forms of high rate photosynthetic intensity found the high rate to be more frequent in the F_1 and F_2 generations than when crossing low intensity forms. The experiences gained with rice have been contrasted (Hayashi 1978, Fig. 12).

Some hybrids exceeded, others remained below the values of the parents. Most hybrids showed intermediary inheritance.

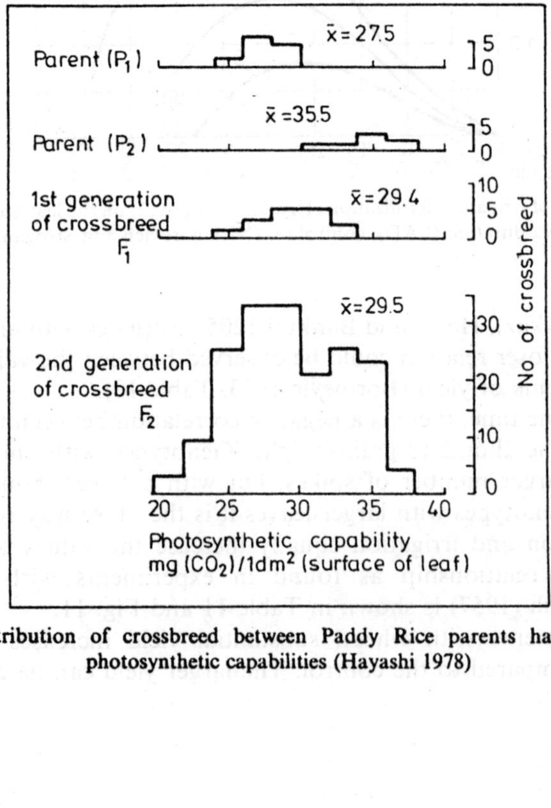

Fig. 12. **Distribution of crossbreed between Paddy Rice parents having different photosynthetic capabilities (Hayashi 1978)**

In an experiment carried out with interspecific hybridization (Björkman et al. 1974) studied the possibility of crossing the more efficient Hatch-Slack C_4CO_2 binding system with the less efficient Calvin-cycle C_3CO_2 binding system, and a way of introducing the (C_4) system into the C_3-type plant species. For informa...on on the segregation conditions of the parent species *Atriplex rosea* (C_4) and *A. patula* and their hybrids refer to Fig. 13. (Björkman et al. 1974).

The figure indicates the dominance of the less efficient pathway and an even lower performance. The properties of the illustrated F_2 hybrids, in agreement with 300 examined hybrids, reveal that there was no distinction with respect to a more efficient system. A more detailed examination of the problem in crosses within species is a task for future research.

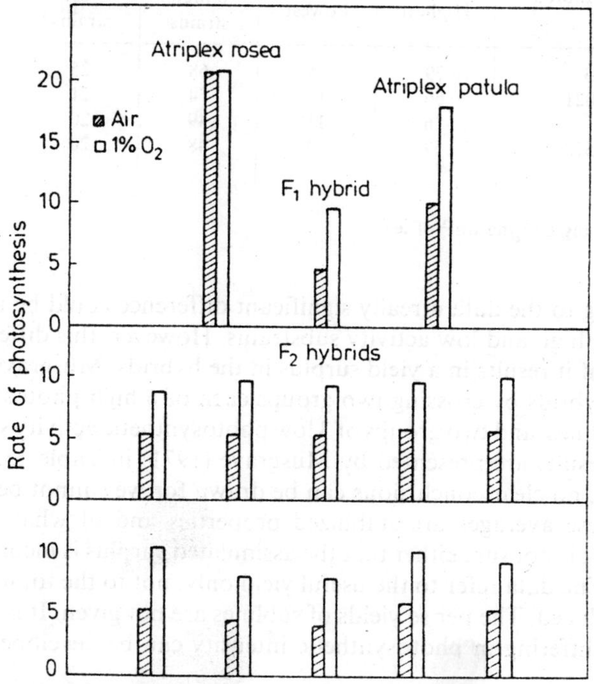

Fig. 13. Light saturated photosynthetic rate in *Atriplex rosea* (C_4) and *A. patula* (C_3) and their F_1 and F_2 hybrids (Björkman et al. 1974). Note: measurements were made at 27 °C, 0.03% CO_2, 21 or 1% O_2 content

The productivity of photosynthesis—as we have seen—can be increased by crossing, but results can only be expected from it when strains of higher photosynthetic intensity are available. Such strains can be produced by selection.

In the course of a breeding programme started in 1967, Musgrave (1971) selected maize sub-strains of higher and lower photosynthetic activity within 4 lines. Their performance is shown by Musgrave (1971) in Table 12.

Table 12

Performance of photosynthetic strains of corn (Musgrave 1971)

Inbred	Rate of photosynthesis*			
	Range		Average	
	Highest	Lowest	6 high strains	3 low strains
C153	79	8	68	25
NY821	66	16	54	21
B8	56	11	49	20
OH51A	57	14	48	20

* mg CO_2/ha/dm^2 of leaf.

According to the data a really significant difference could be attained between the high and low activity substrains. However, this difference is only useful if it results in a yield surplus in the hybrids. Musgrave (1971) produced hybrids by crossing two groups each of 6 high photosynthetic activity sublines, and two groups of 3 low photosynthetic activity sublines. The yield results are presented by Musgrave (1971) in Table 13.

However, no clear conclusions can be drawn for we cannot be sure of whether these averages are stabilized properties and of what their h^2 values are. It is not sure either that the assimilated surplus is incorporated in the ear. The data refer to the useful yield only, not to the total organic matter produced. The per se yields of sublines are not given. It is obvious that types differing in photosynthetic intensity can be developed in the lines.

Types showing different photosynthetic intensity can be produced by *mutagenic treatment* too, as pointed out in experiments conducted by Yamaguchi et al. (1976) with rice. They produced semi-dwarf mutants (G

26, G 29, G 31 and G 32) in the varieties NF-1 and Meitoku 5, with two control varieties included in the experiment.

The flag-leaf and the two upper leaves had 40–50 per cent smaller surfaces in the mutants than in the control variety Sasanishiki. In spite of this the former exceeded the latter in the amount of dry matter/dm², total and protein N contents (Yamaguchi et al. 1976, Table 14, Fig. 14).

As for the intensity of photosynthesis the mutant was superior not only to the variety Sasanishiki but also to the parent varieties. It is interesting that the authors found the rate of photosynthesis per unit leaf surface (dm²) to have a correlation of $r=0.9097$ at 1% level with the dry matter content of the leaf, and of $r=+0.9325$ with the N content.

The leaf area is an important concomitant in increasing photosynthetic yield. By increasing the leaf area (stand density) an about threefold

Table 13

Yields of single-crosses of photosynthetic strains of corn (Musgrave 1971)

Single-cross ranking for photosynthesis	NY821 × C153 bu/A*	OH51A × B8 bu/A*
H-1	170	170
H-2	162	134
H-3	146	—
H-4	166	150
H-5	168	158
H-6	166	130
L-7	162	154
L-8	181	134
L-9	154	134
PB	146	152

* Average of 4 repetations.

Table 14

Varietal differences in the dry matter and the nitrogen contents per unit leaf area (dm²) (Yamaguchi et al. 1976)

	Sasanishiki	Reimei	Meitoku 5	NF-1	G 26	G 29	G 31	G 32
Dry matter (g)	0.467	0.502	0.535	0.654	0.794	0.763	0.699	0.792
Total nitrogen (mg)	11.5	15.2	15.1	25.3	26.6	22.1	26.2	28.9
Protein-N (mg)	10.2	13.9	13.8	22.9	22.9	19.2	22.9	26.8

photosynthetic productivity compared to the 2.18 per cent rate of light utilization in maize has been attained with perennial grasses in Holland.

The leaf area (L) has two components: the number of leaves and the size of the leaf. Duarte and Adams (1963 in: Wallace et al. 1972) crossed a bean variety with few large leaves and one with many small leaves. In the F_1 generation the high leaf-number was dominant, the size of leaf showed intermediary inheritance. The latter characteristic was probably determined by additive genes. Fowler and Rasmusson (1969 ibid.) studied the inheritance of the flag-leaf and the two adjoining lower leaves in barley. In the F_1 L was intermediate between the values of the parents. The h^2 values obtained from the regression of the F_4 plants compared to the F_3 generation ranged between 0.18 and 0.73 in the different combinations.

Edward (1970 in: Wallace et al. 1972) selected 4 lines from a *Lolium perenne* population. They were characterized by large and small leaves, and the rapid and slow rate of appearance of leaves, respectively. By comparing the F_1 and F_2 generations the author established that the rate of appearance of leaves and the elongation of the individual leaves are influenced by additive genes. Non-additive inheritance was pointed out for the total leaf area and its components: the length and width of the leaf. The rate of development of leaf area and leaf dimensions showed expressed heterosis.

Selection for larger leaves or faster rate of development of new leaves involved the deterioration of other properties. The vascular system that ensures the rapid growth of a leaf at the same time inhibits the development of the lower leaves. In *Lolium multiflorium* the h^2 value of length, width and leaf area ranged between 40 and 70 per cent. Lupton et al. (1967) examined the h^2 values of L and its components in wheat, and found that the leading role in determining the leaf area was played by additive gene actions. Kraljevic-Balalic (1973) drew similar conclusions from the results of an experiment with 4 wheat varieties, in which the h^2 value of the mentioned property ranged from 6 to 47 per cent. In crossing combinations, superdominance causing a heterosis effect frequently occurs. One-way selection has a negative effect on the other components in this case too.

Investigations on a number of plant species (rice, soybean, batata, etc.) have unequivocally proved (Wallace et al. 1972) that the photosynthetic activity in a thicker leaf (especially if it consists of many tiny cells) is higher than in a thinner one. It decreases the so-called mesophyll resistance, that is the resistance within the leaf to CO_2 flowing to the chloroplasts, and increases the CO_2 fixation of the carboxylating enzymes, since the number of enzymes per unit leaf area is larger.

With a view to a possible increase in stand density many varieties of *upright leaf* have recently been produced. Hayashi and Ito (1962), Japanese researchers, were the first to point out that rice varieties with erect leaves developed better and gave larger yields than the normal types. After the rice they obtained similar experiences with wheat, barley, soya and sugar-beet. In maize with a genetically analogous hybrid, and in the hybrid Pioneer 3306 by the mechanical adjustment of the angle of inclination of leaves, Pendleton (1968) attained interesting results (Table 15).

In the genetically analogous hybrid the proportion of sterile plants was substantially reduced (at a high density of the stand).

In maize the liguleless gene causes an upright position of leaves.

Piovaric and Vidovic (1973) introduced the lg_1, lg_2 and lg_3 genes into lines of the single crosshybrid Oh 43 × 014 and tested the isogenic hybrids with a plant number of 41,000, 55,000 and 80,000/ha, respectively. The effects of the three genes were similar, and in accordance with the correlation seen in Table 15 a 13 per cent yield surplus occurred in the 80,000 plant/ha stand.

Further favourable results have been obtained with wheat and rice; work done with maize has not been as successful as that.

Tolerance to increased stand density can be brought about by selection, without changing the angle of inclination of leaves. In lines produced from synthetic varieties (F and M) Tsotsis (1972) and his collaborators have applied the principle of gradually increasing stress. This means that they increased the initial plant number of 57,200/ha (26,000/acre) by 8,800 every year until they reached a stand density of 111,000 plant/ha (Tsotsis 1972, Fig. 14).

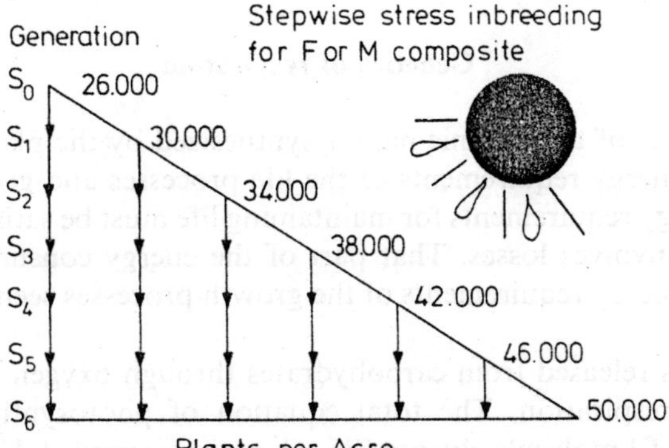

Fig. 14. Stepwise inbreeding (Tsotsis 1972)

They wanted to increase the present 3.0–3.5 leaf area index of hybrids to a value of 4.5–6.0. In comparison to the lines the LAI of hybrids increases which may exercise a favourable influence on the yield.

Table 15

Grain yields and barren plants from leaf angle study (Pendleton 1968)

Comparisons	Yield*	Plants barren* (%)
Genetic isolines of hybrid C 103 × Hy		
1 Normal leaf	99 a	28 a
2 Upright leaf	140 b	14 b
Mechanical manipulation of leaf angle of Pioneer 3306		
3 Normal (untreated)	170 c	4 c
4 All leaves positioned upright	181 cd	6 bc
5 Leaves above ear positioned upright	194 d	3 c

* Means with the same letter are not significantly different at 5% level.

***

The investigations described above show that there are great possibilities of increasing the productivity of photosynthesis by altering the genetic and environmental factors. Loss is another part of the problem; respiration, as the most important of its factors will be dealt with first.

Genetics of respiration

A major part of the organic matter synthesized by the plants is used to cover the energy requirements of the life processes and growth.

The energy requirements for maintaining life must be satisfied, and this inevitably involves losses. That part of the energy consumption which meets the energy requirements of the growth processes seems to be more useful.

Energy is released from carbohydrates through oxygen uptake in the course of respiration. The total equation of photosynthesis: for the formation of 1 molecule glucose 675 cal. energy is needed. From 1 glucose molecule 304 calories are released and are bound in the course of respiration; the complete oxidation of glucose means a free energy reduction of $\Delta G^0 = -675$ cal/mol, the efficiency is 45 per cent (Szalai 1974). This is realized in a reaction chain evolved during the phylogenesis of plants and cannot be changed (James 1965 in: Szalai 1974, Table 16).

The individual species and varieties, however show differences depending on the phases of ontogenesis (Table 17 from Péterfi 1956, in: Szalai 1974) concerning the intensity of respiration, as for what percentage

Table 16

ml CO_2 per mg dry matter discharged in one hour (James 1965, in: Szalai 1974)

The respirating organ or organism	QCO_2
Baking yeast	60–100
Aspergillus niger, 2-day culture	78
Arum italicum spedix	31
Solanum lycopersicum, root tip	6–8
Hordeum, seedlings	1.6
Pisum, dry seeds	0.00012

Table 17

CO_2 discharge of sunflower leaves of various age (Péterfi 1956, in: Szalai 1974)

Sunflower	CO_2 discharged/h for 1 g dry weight, mg
22 day-old leaves	3.00
36 day-old leaves	0.81
50 day-old leaves	4.46
64 day-old leaves	0.34
136 day-old leaves	0.08

of the respiration is utilized in the growth processes (Table 18, Ishizuka 1969), and in the presence of the two forms of respiration: light and dark respiration. The trend of the latter in cotton as a function of temperature is shown in Fig. 15 (Baker et al. 1972).

From the point of view of organic matter production the 32/23 °C variant is the most favourable.

Table 18

Growth efficiency of rice plants in a community (Ishizuka 1969)

Stage of growth	Growth efficiency, (%)	Percentage of respiration directly related to growth
Early stage	(60)	(100)
From booting stage to flowering	41	68
From flowering to milky stage	36	60
From milky stage to complete ripening	27	44

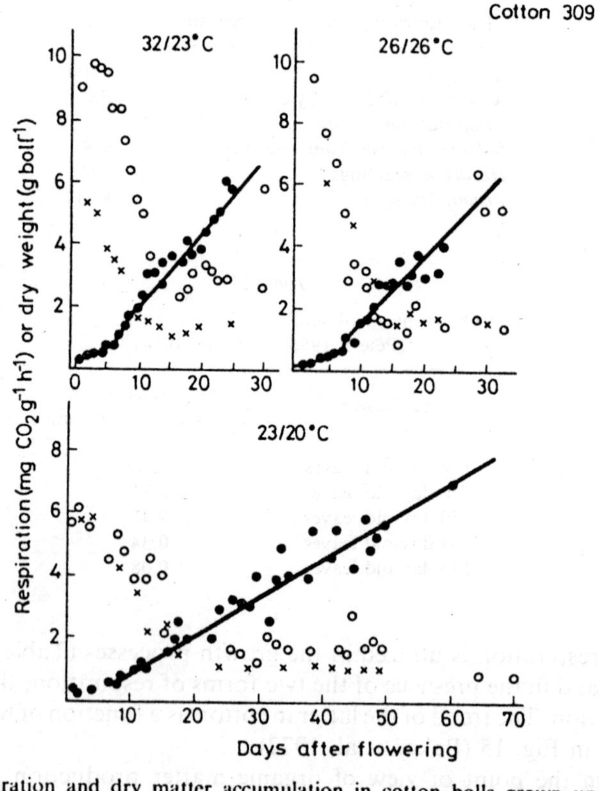

Fig. 15. Respiration and dry matter accumulation in cotton bolls grown under three temperature regimes. Respiration day–o; respiration night–x; dry weight (Baker et al. 1972)

In the course of respiration the plants discharge 5 to 10 times as much CO_2 as their own volume, and about one-third to one fifth of the amount is fixed during photosynthesis. The C_4 plants can utilize part of this carbohydrate; explaining why their net rate of assimilation is more favourable (Szalai 1974, Hess 1968).

Regarding respiration losses, those types in which photorespiration does not take place are, naturally, more favourable. Apart from the highly productive tropical grasses, this group includes maize, *Sorghum bicolor*, and of the dicotyledons, the genus *Amaranthus*. Taking these into consideration, 300 varieties of leguminous species—including the soya (Koo 1972)—were examined to find types in which photorespiration does not take place. Since this has failed, the author has tried to select such a type from among induced mutants.

There may be differences between the varieties in the rate of dark respiration too. In the bean variety M62, the proportion of dark respiration is higher than in the varieties Red Kidney and Redkote (Izhar and Wallace 1967, in: Wallace et al. 1972). Heichel (1971, in: Wallace et al. 1972) found a lower rate of dark respiration and more rapid accumulation of organic matter in one of two inbred maize lines than in the other. This is due to the fact that this line used up less organic matter for respiration and could spend more for building up the organs. The loss is 22 and 33 per cent of the synthesized organic matter, respectively. Having summed up the data of various authors Zelitsh (1971) found this loss to be 16–29 per cent in maize, 27–48 per cent in wheat and 33–70 per cent in lucerne. The energy released by dark respiration may be connected with the formation of ATP, NADM or NADPH and thereby with growth, but there are forms of energy release that are in no connection with these processes.

The optimum of organic matter production depends on the level and difference between the rates of assimilation and dissimilation (photosynthesis and respiration). Differences in temperature and light demands between the plant species are well characterized by the data presented in Tables 19 and 20 (Samokhvalov 1972).

Keeping in mind the above, and reckoning with the uncontrollable meteorological factors, we have to try to attain the maximum of biological yield (organic matter production) with the varieties and stand densities chosen.

Table 19

Effect of temperature on the intensity of photosynthesis and respiration (mg CO_2/g dry matter of leaf/h) (Samokhvalov 1972)

Species		CO_2 mg temperature		
		12	20	36
Maize	P	2.32	17.21	19.08
	R	0.36	5.17	5.59
	D	1.96	12.04	13.49
Carthamus tinctorius	P	10.04	20.99	22.25
	R	4.36	8.49	8.73
	D	6.68	12.50	13.52
Oat	P	5.66	34.97	19.72
	R	1.33	10.05	6.10
	D	4.33	24.92	13.62
Fibre flax	P	7.46	17.80	5.68
	R	2.38	5.88	1.64
	D	5.08	11.92	4.04

P = photosynthesis, R = respiration, D = difference.

Table 20

Effect of light intensity on photosynthesis and respiration (mg CO_2/g dry matter of leaf/h) (Samokhvalov 1972)

Species		Illumination, 1000 lux			
		6	12	16	24
Maize	P	6.00	8.32	9.20	9.07
	R	0.66	1.74	1.92	1.88
	D	5.34	6.58	7.28	7.19
Carthamus tinctorius	P	6.60	10.50	12.80	12.65
	R	1.34	2.32	2.86	2.72
	D	5.26	8.18	9.94	9.83
Oat	P	9.66	10.90	8.91	8.15
	R	2.26	2.93	2.51	2.10
	D	7.40	7.97	6.40	6.05
Fibre flax	P	13.07	13.12	11.31	9.06
	R	1.36	1.49	1.27	1.06
	D	11.71	11.63	10.04	8.00

P = photosynthesis, R = respiration, D = difference.

Growth, development, earliness

The so-called net rate of assimilation is (NAR) namely the difference between the organic matter synthesized in the course of assimilation and that spent during respiration. The correlation between the net assimilation rate and the leaf area is expressed by the assimilation performance (AP):

$$AP = \frac{\text{dry matter increment g/m}^2 \times \text{day}}{\text{average leaf area} \times \text{number of days}}$$

The NAR varies with species, variety and even plant. Watson (1952 in: Bálint 1976) found the average NAR in wheat, barley and oat to be 0.350, 0.312 and 0.287, respectively, per week. In cacao seedlings originating from 4 different trees, Goodall (1950 ibid.) observed, steadily transmitted NAR differences. According to Boonstra (1939 ibid.) in high sugar content beet varieties NAR is higher than in those containing less sugar.

The literature has replaced the NAR by one of its elements when characterizing the process. This element is the net carbohydrate exchange (NCE) which can be easily studied by infrared CO_2 analysis.

The value of NCE in the bean variety Michelite-62 is 9–31 per cent higher than in 4 other varieties; 25 per cent higher than in the Red Kidney. In 10 sugar-beet varieties, a maximum of 2.5-fold difference was measured; in soybean varieties the difference was 48 per cent, while in 27 maize lines a threefold difference could be observed.

The value of NCE was higher by 26–36 per cent in the dwarf wheat variety Mexipak than in the variety C273. In wheat varieties a rise in the LAI reduced the value of NCE (Wallace et al. 1972).

The situation is similar in maize (Voldeng 1971, Fig. 16), but owing to the higher LAI the crop growth rate (CGR) per m^2 will be higher, particularly at the beginning of development.

The rising value of LAI—as we have seen—increases the crop yield only to a certain extent. In an experiment aimed at reaching maximum yields of rice Murata and Togari (1972) found a negative correlation between NAR and LAI (Table 21).

The reason for the negative correlation may be that the increased stand density aimed at obtaining large yields reduces the NAR value because of the greater loss of respiration (Fig. 9).

The growth rate is only one of the factors of plant productivity. Here too, like in all characteristics there are great differences between the species, as reported by Evans (1972 in: Evans 1975) (Table 22).

Investigations conducted for several weeks did not prove the superiority of C_4 plants to the C_3 type. This may be due to the fact that at a lower level of illumination the photosynthetic intensity of C_3 plants is higher, under short day conditions the dark respiration loss of C_4 plants increases.

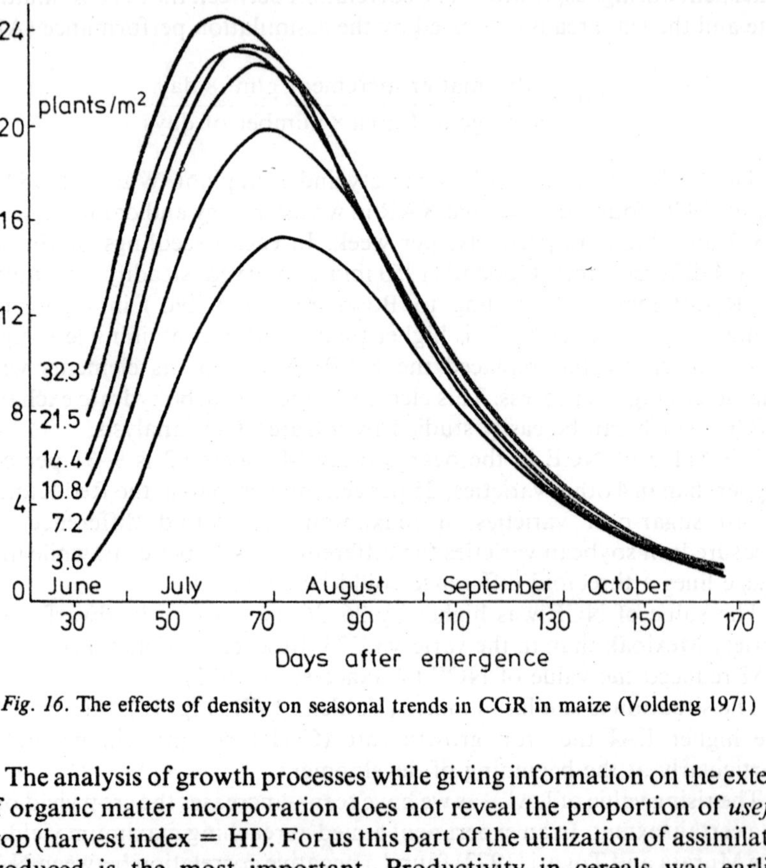

Fig. 16. The effects of density on seasonal trends in CGR in maize (Voldeng 1971)

The analysis of growth processes while giving information on the extent of organic matter incorporation does not reveal the proportion of *useful* crop (harvest index = HI). For us this part of the utilization of assimilates produced is the most important. Productivity in cereals was earlier characterized by the ratio of grain to straw. In Hungary the old varieties, e.g. B 1201, were characterized by a grain/straw ratio of 1 : 2.5. According to Balla and Szunics (1978) in the current varieties this ratio is 1 : 1.5–1.7. In one of the present English varieties (Bingham 1972) the ratio of grain to straw is 1 : 0.62. This variety (TL 365/25) is 72 cm high compared to the 113 cm plant height and 1 : 1.1 grain/straw ratio of earlier varieties (Maris Widgeon).

The grain/straw ratio has supplied useful data for the work of breeding, but the principle of developing types of higher physiological economy has not become a main objective of breeding. There have been but few data available on changes in this index under different ecological conditions, although in deciding the application of various *cultural practices* (fertilizing, changing the sowing time, increasing the stand density, irrigating, etc.) this index is of great help.

Before presenting examples of crops grown for their grain, let us mention a flower in this context. Some Gerbera plants may even bring

Table 21

Correlation coefficient of grain yield (1) with LAI at heading (2) and with the number of spikelets (3) in the IBP 'Maximal growth rate experiment' (Murata and Togari 1972)

Year	$r_{1.2}$	$r_{1.3}$	$r_{1.2.3}$
1967	0.600⁺	0.765⁺⁺⁺	−0.033
1968	0.787⁺⁺	0.858⁺⁺	0.579
1969	0.260	0.554	0.221
Pooled	0.492⁺⁺	0.683⁺⁺	0.203

⁺ Significant at 5%, ⁺⁺ at 1% and ⁺⁺⁺ at 0.1% level.

Table 22

Maximum recorded short term crop growth rates (CGR) (Evans 1975)

Species	Photo-synthetic pathway	CGR (g m^{-2} ground day^{-1})	Reference
Helianthus annuus	C_3	68	Hiroi and Monsi 1966
Pennisetum purpureum	C_4	60	Arias and Butterworth 1965
Agrostemma githago	C_3	57	Blackman 1962
Phragmites communis	C_3	57	Dykyjova 1971
Oryza sativa	C_3	55	Tanaka et. al. 1966
Pennisetum typhoides	C_4	54	Begg 1965
Typha latifolia	C_3	53	Williams et al. 1965
Zea mays	C_4	52	Williams et al. 1965
Sorghum vulgare	C_4	51	Loomis and Williams 1963
Saccharum officinarum	C_4	38	Borden 1942
Solanum tuberosum	C_3	37	Lorenz 1944
Beta maritima	C_3	31	Watson 1958
Ananas comosus	CAM	28	Eckern 1964

4–10 flowers each. A glasshouse surface of 1 m^2 is obviously better utilized with a variety that produces 8–10 instead of 4–5 flowers per plant, redoubling thereby the value of HI (Szegedi 1979, personal communication).

In an experiment with bean plants Wallace et al. (1972) demonstrated the favourable effect of increasing the value of HI in identical types (Wallace et al. 1972, Table 23).

According to the data of Győrffy et al. (1965) in 1962 the proportion of grain crop at Martonvásár was 36.4 in the open pollinating varieties and 45.8 per cent in the Martonvásár maize hybrids. In the case of optimum plant number and sowing time this HI value has been found by Menyhért and Ángyán (1979, personal communication) to be 0.50, 0.47 and 0.40 in the hybrids KSC 360, SZTC-344 and MVSC-580, respectively. The values obtained by Hajós (1979, personal communication) with hybrids of various origin and vegetation period were even more favourable, particularly those of two FAO-200 combinations sown to a stand density of 50,000 plant per ha (0.54, 0.51).

Since 1962 the actual value of HI has not shown much improvement, here and there even a set-back has occurred, particularly with hybrids of higher FAO-numbers. But that is not the point; progress is seen in the fact that the HI-values of good hybrids remain even with *substantially increasing stand densities,* whereby considerably higher yield averages can be attained. This statement is supported by the data of Fig. 17—except those for the hybrid MVSC-580—obtained in experiments conducted by Menyhért and Ángyán (1979).

Table 23

Relative seed yields of seven dry bean varieties as determined over 6 years from 18–20 yield trials (Wallace et al. 1972)

Class of dry bean	Variety	Relative yield	Harvest index (%)	Yield advantage (%)
Marrow	Perry Marrow	100	60	
	Cornell 7–16	121	64	21
Yellow eye	Yellow eye	93	60	
	Steuben	111	62	20
Pea	Michelite	111	55	
	Monroe	121	57	9
Red kidney	Red Kidney	108	57	

These correlations were studied by DeLoughery and Crookston (1979) with more extreme stand densities, under normal and stress conditions (drought), with ten hybrids belonging to 5 Minnesota maturity group systems.

The results are shown by DeLoughery and Crookston (1979) in Tables 24 and 25 and Figs 18 and 19.

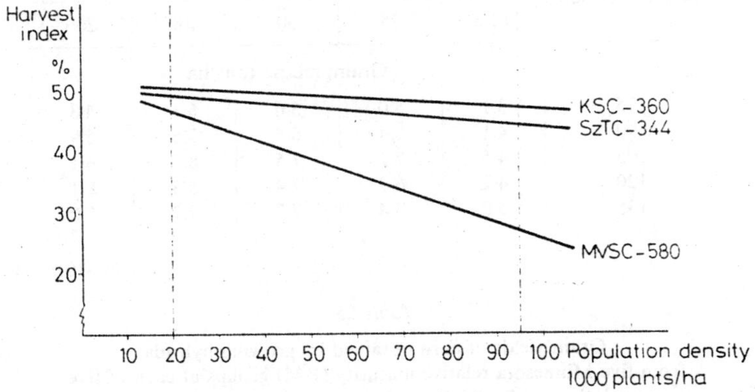

Fig. 17. Relationship between plant number and harvest index in three hybrids (Menyhért and Ángyán 1979, personal communication)

The data truly represent the effects of environmental factors (stand density, dry weather) on HI.

In an average stand density (50,000 plant/ha) the adapted hybrids (relative maturity group RM 105) showed an average HI value of 0.39 (ranging from 0.32 to 0.47) irrespective of the environment. The highest HI value in all places was 0.5.

The growing number of plants greatly decreased the HI. With a plant number of 100,000/ha or more the plants practically produced leaves and stalks alone. Under dry conditions it would have been useful to put several intermediate plant numbers in between 50,000 and 100,000. It is anyway obvious that the decrease of HI due to increasing stand densities of early hybrids is of lower extent under dry conditions. Under normal conditions hybrids of different maturity groups give rather similar responses. Any considerable reduction in the value of HI has been observed only with a plant number of 200,000/ha. It has been established that in a state of stress there is a close positive correlation between HI and grain yield (r=0.96),

Table 24

Grain yields of corn obtained by growing hybrids from five Minnesota relative maturity (RM) groups at each of five populations at Waseca (nonstress environment). Each value is a mean based on six observations (SD = 0.7) (DeLoughery and Crookston 1979)

RM of hybrids	Population (plants/ha × 1,000)				
	12.5	25	50	100	200
	Grain, metric tons/ha				
75	3.1	5.0	5.6	6.1	4.3
90	4.1	5.4	6.8	7.6	5.9
105	4.5	7.2	7.5	8.7	6.1
120	4.2	6.9	7.4	8.5	8.1
135	5.0	7.4	7.7	7.7	7.0

Table 25

Grain yields of corn obtained by growing hybrids from five Minnesota relative maturity (RM) groups at each of five populations at St. Paul (stress environment). Each value is a mean based on six observations (SD = 0.2) (DeLoughery and Crookston 1979)

RM of hybrids	Population (plants/ha × 1,000)				
	12.5	25	50	100	200
	Grain, metric tons/ha				
75	3.2	3.3	3.2	1.5	1.1
90	3.9	3.7	3.7	2.2	1.2
105	3.7	3.5	2.5	0.9	0.7
120	3.8	3.3	2.1	0.7	0.6
135	3.2	2.9	1.3	0.1	0.0

while under normal circumstances this correlation is weak ($r = 0.28$). This suggests that from the HI value determined for a site no conclusion can be drawn on the proportion of grain yield under different environmental conditions.

Knowing the HI value, optimum production technologies can be elaborated. Sowing time is also an important factor, as pointed out in an experiment series conducted after Johnson and Major (1979), in which 10

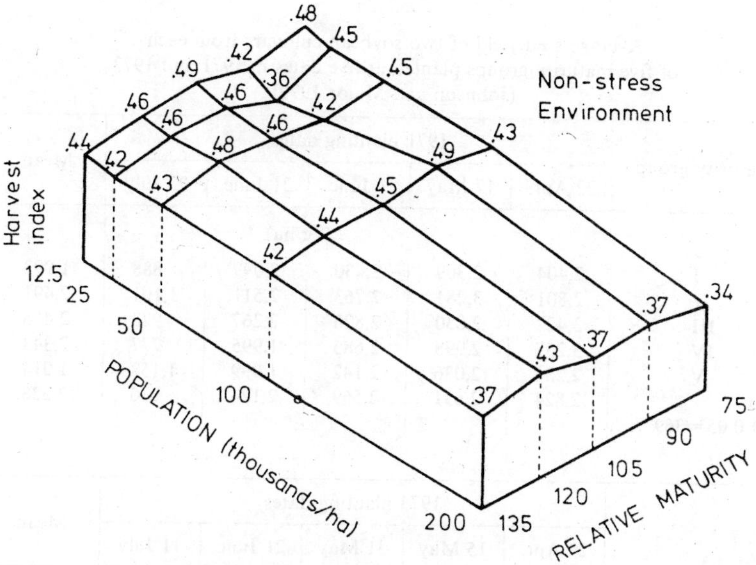

Fig. 18. Harvest index as affected by population and relative maturity at Waseca (nonstress environment). Each value is a mean based on six observations ($S_D = 0.03$) (DeLoughery and Crookston 1979)

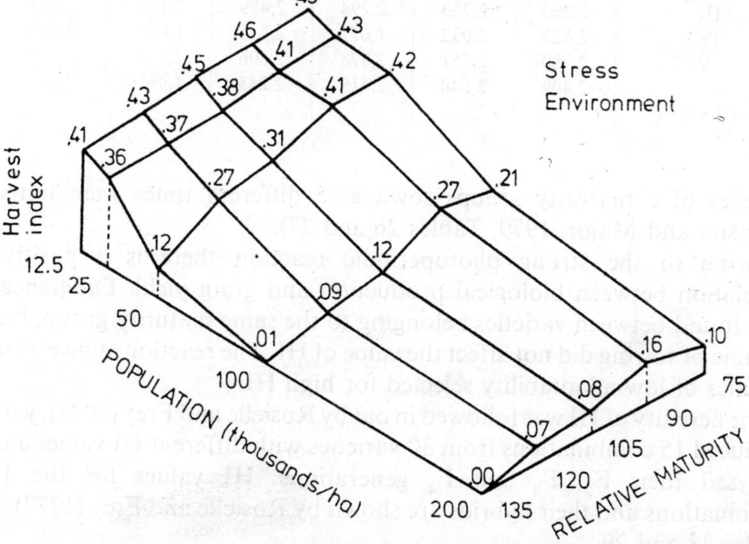

Fig. 19. Harvest index as affected by population and relative maturity at St. Paul (stress environment). Each value is a mean based on six observations ($S_D = 0.03$) (DeLoughery and Crookston 1979)

Table 26

Average seed yield of two soybean cultivars from each of five maturity groups planted at five dates in 1971 and 1973 (Johnson and Major 1979)

Maturity group	1971 planting dates					Mean
	22 Apr.	17 May	1 June	21 June	22 July	
	(kg/ha)					
I	2,404	2,309	2,430	2,099	888	1,992
II	2,801	3,281	2,763	2,511	1,101	2,491
III	3,422	2,830	2,826	2,267	928	2,455
IV	3,257	2,998	2,685	1,995	777	2,343
V	2,234	2,036	2,142	1,999	1,158	1,914
Mean	2,824	2,691	2,569	2,174	930	2,238
LSD 0.05 = 369						

	1973 planting dates					Mean
	29 Apr.	15 May	31 May	21 June	11 July	
	(kg/ha)					
I	2,034	2,242	2,476	1,530	1,696	1,996
II	2,345	2,964	2,773	2,065	1,893	2,408
III	2,260	2,753	2,294	2,479	2,136	2,384
IV	2,623	2,032	1,671	2,345	1,656	2,065
V	2,780	2,751	2,878	2,806	2,372	2,717
Mean	2,409	2,548	2,419	2,245	1,951	2,315
LSD 0.05 = 0.04						

varieties of 5 maturity groups sown at 5 different times were tested (Johnson and Major, 1979, Tables 26 and 27).

Owing to the strong photoperiodic reaction there is a positive correlation between biological production and grain yield. Differences were found between varieties belonging to the same maturity group, but the time of sowing did not affect the value of HI. The reaction appeared in varieties of low adaptability selected for high HI.

The heredity of HI was followed in oat by Rosielle and Frey (1977), who produced 15 combinations from 30 varieties with different HI values and analysed their F_2, F_3 and F_4 generations. HI values for the 15 combinations and their hybrids are shown by Rosielle and Frey (1977) in Tables 28 and 29.

According to experimental results additive gene effects decisively determine HI. In combinations where both parents have high HI values the phenotypic variation of HI and other properties is low.

In connection with the increase of HI some physiological correlations are worth examining. One of them is the *effect of fruit formation on the performance of assimilation.* In the literature it is referred to as "source-sink relationship".

Bingham (1972) studied the phenomenon in a wheat "field crop of Cappelle-Desprez by emasculating ears and pollinating different numbers of florets, in order to obtain treatments with a constant photosynthetic capacity but differing in sink size. With each reduction in grain number, the weight of individual grains at corresponding floret positions increased, from 56.5 mg in ears with 41 grains to 68.5 mg in those with 16. This implies that in untreated ears the grain did not reach any upper limit in size

Table 27

Average harvest index of two soybean cultivars from each of five maturity groups planted at five dates in 1971 and 1973 (Johnson and Major 1979)

Maturity group	1971 planting dates					Mean
	22 Apr.	17 May	1 June	21 June	22 July	
I	0.37	0.34	0.37	0.37	0.31	0.35
II	0.39	0.38	0.37	0.41	0.35	0.38
III	0.34	0.33	0.34	0.35	0.27^{+}	0.33
IV	0.35	0.33	0.33	0.34	0.29^{+}	0.33
V	0.25^{+}	0.26^{+}	0.26^{+}	0.24^{+}	0.27^{+}	0.26
Mean	0.34	0.33	0.33	0.34	0.30	0.33
LSD 0.05 = 0.14						

Maturity group	1973 planting dates					Mean
	29 Apr.	15 May	31 May	21 June	11 July	
I	0.39	0.36	0.37	0.29	0.38	0.36
II	0.41	0.40	0.36	0.35	0.40	0.38
III	0.34	0.33	0.27	0.35	0.35	0.33
IV	0.33	0.25	0.25	0.34	0.30	0.29
V	0.35	0.31	0.33	0.36	0.36	0.34
Mean	0.36	0.33	0.32	0.34	0.36	0.34
LSD 0.05 = 0.05						

$^{+}$ Killed by frost.

4*

which might be due to a limited capacity of the grain to store carbohydrates, thus indicating a dependence of yield on photosynthetic capacity. However, the increases in grain size were inadequate for compensating completely in total weight of grain per ear (which fell from 2.1 g to 1.0 g per ear) for the same treatments, so the ability of the ear to store carbohydrate was also important."

Table 28

Means of six traits measures on ten check cultivars oats. (Weighted averages from three experiments)⁺ (Rosielle and Frey 1977)

Cultivar name	Trait					
	Heading date (days)	Plant height (cm)	Plant weight (q/ha)	Grain yield (q/ha)	100-seed weight (g)	Harvest index (%)
O'Brien	21.1	102	71	32	2.89	45.9
Grundy	18.8	91	71	34	2.91	47.5
X434-II	16.6	95	73	36	2.90	48.9
Clintford	18.7	91	74	34	3.12	45.6
Pettis	17.0	99	73	35	2.46	48.4
Otee	20.8	91	73	32	2.70	43.2
Tippecanoe	19.0	94	72	29	2.81	40.2
Garland	22.0	94	67	30	2.77	44.2
Portal	22.4	101	75	33	2.78	45.4
Dal	24.8	98	70	28	2.65	39.4
Standard error	0.2	0.5	1.3	0.6	0.02	0.3

⁺ For all traits except 100-seed weight, data from the 1974, early-planted experiment was given twice the weight in calculating these means so they may be compared with statistics calculated from overall line means.

Having removed the upper third and two-third of the spike in the Pilot variety, Damisch (1973) found the thousand-grain-weight to increase by 26 and 17 per cent, with a parallel 6 and 31 per cent reduction in the yield, respectively. Similar tendencies were observed in the varieties Pluto and Salzmünder Bartweizen. In the varieties Kavkaz and Panoniya the thousand-grain-weight increased by 2 and 4 per cent, respectively, after the removal of one-third of the spike, and decreased by 6 and 2 per cent, respectively, when two-thirds of the spike was removed. The yield reduction was 17 and 31 per cent in the Kavkaz, and 22 and 20 per cent in the variety Panoniya, respectively. The responses of the varieties were thus rather diverse.

Table 29

Deviations of cross from midparental means for plant weight, grain yield, and harvest index in the 74 F_3 and 74 F_4 data sets and deviations of F_3 from $F_4(F_3–F_4)$. (Expressed as percentages of midparental means) (Rosielle and Frey 1977)

Cross group	Cross	Plant weight			Grain yield			Harvest index		
		F_3	F_4	$F_3–F_4$	F_3	F_4	$F_3–F_4$	F_3	F_4	$F_3–F_4$
High × Low	1	6.1**	7.6**	−1.5	9.4**	9.4**	0.0	3.2**	2.0	1.2*
	2	5.6*	−2.6	8.2**	9.3**	−0.1	9.4**	3.2**	2.3*	0.9
	3	16.5**	9.3**	7.3**	19.9**	11.5**	8.4**	2.9**	2.0	0.9
	4	16.2**	7.2**	9.0**	16.8**	6.4*	10.4**	0.6	−0.8	1.4*
	5	6.2*	4.1	2.1	4.4	0.7	3.7*	−0.9	−2.3*	1.5*
	6	13.0**	7.2**	5.7**	14.6**	8.2**	6.4**	1.5	0.9	0.6
High × High	7	9.2**	8.1**	1.1	9.3**	7.6**	1.7	0.0	−0.6	0.6
	8	−0.8	0.7	−1.5	1.1	3.4	−2.3	2.2*	2.8**	−0.6
	9	9.3**	7.7**	1.6	9.3**	6.5**	2.8*	−0.1	−1.2	1.1*
	10	−0.7	−0.3	−0.4	−0.3	0.6	−0.9	0.2	0.7	−0.5
	11	8.6**	4.4	4.2**	6.5*	2.3	4.2**	−1.7	−2.0*	0.3
	12	2.7	4.3	−1.6	1.8	4.4	−2.6*	−0.7	0.1	−0.8
Low × Low	13	11.3**	7.5**	3.8*	13.4**	9.6**	3.8*	2.5*	2.4*	0.1
	14	7.1*	3.0	4.1*	11.2**	5.1	6.1**	4.2**	2.5*	1.7*
	15	10.0**	7.1**	2.9*	10.5**	6.1*	4.4**	0.3	−1.0	1.3*

*,** Significant at the 5 and 1% levels, respectively.

Damisch (1973) also examined the effect of reducing the surface of assimilation. In the same varieties the flag-leaf was reduced to half or removed completely. In the Pilot variety both treatments resulted in a 9 per cent yield reduction, while in Kavkaz this value was 16 per cent. In other varieties the removal of half of the flag-leaf caused no reliable change in the yield, while in response to the total removal of the flag-leaf it decreased by 2, 21 and 23 per cent, respectively. Again, the varieties did not give uniform responses.

In the potato the removal of the tubers or a slowing-down in their growth rate due to lower soil temperatures resulted in reduced net assimilation rate and photosynthesis (Burt 1965), while the partial removal of leaves did not cause any change in the growth rate of tubers (Humphries 1969).

Shannon (1978) studied the correlation in maize and arrived at the following conclusion: in regions with short vegetation periods the capacity of photosynthesis set limits to the yield. On such areas—e.g. in Canada—besides increasing the LAI, efforts should be made to improve the intensity of photosynthesis; the possibility to do so has been testified in the experiments of Crosby et al. (1977 in: Shannon 1978). In the mid-western maize zone of the United States of America and in Mexico the source is satisfactory; it is the "sink" that controls the productivity there. The "sink" can be regulated by the number of grain/plant, or by the capacity of the "sink" itself. The supply level with assimilates plays the

Table 30

Percentage of assimilates
in wheat grains (Lupton 1966)

Varieties	1960	1961
Capelle	31.8	48.5
Heine VII	46.3	56.2
Holdfast	47.3	69.3
Peko	60.1	77.8
SD (P=0.05)	7.1	9.3

most important role in the period of flowering and early grain development. The capacity of "sink" means the extent of "attraction" exercised by the developing grain on the assimilates, and the rate of their utilization for the production of organic matter, first of all starch. Referring to Stoy (1969) the author notes that in many plants with a

Table 31

Percentage of assimilated carbohydrate translocated to the grain at successive ages (Lupton 1972)

Varieties	Before anthesis	Days after anthesis			
		0–10	11–20	21–30	31–35
Capelle Desprez					
Flag leaves	5.3 ± 0.9	17.0 ± 6.2	47.5 ± 3.1	47.9 ± 7.7	47.2 ± 8.6
Second leaves	2.5 ± 1.1	14.9 ± 1.0	12.0 ± 4.2	8.0 ± 4.0	trace
Ears	trace	21.2 ± 4.8	36.7 ± 5.2	31.5 ± 3.8	
TL 365a/25					
Flag leaves	5.4 ± 0.7	29.0 ± 5.7	49.6 ± 15.2	36.5 ± 6.4	49.2 ± 8.0
Second leaves	2.5 ± 1.2	6.2 ± 3.7	5.9 ± 3.6	7.4 ± 0.9	trace
Ears	trace	26.6 ± 3.8	35.5 ± 11.5	45.2 ± 1.0	

decrease or increase in the metabolic requirements of the "sink" tissues a simultaneous decrease or increase in the photosynthetic intensity of "source" leaves can be observed. As opposed to this thesis, Bingham (1972)—as pointed out before—did not obtain a similarly clear correlation for wheat.

A quite direct correlation between NAR and productivity cannot be established if only because different varieties use different proportions of assimilates for seed development. Lupton (in: Bálint 1977) used radioactive carbon dioxide to follow the translocation of assimilates in wheat. Tables 30, 31 (Lupton 1972) show what proportions of the carbon compounds present in the grain originate from the spike, and what proportions from the assimilation products of other plant parts.

In a later experiment—after a thorough analysis of the development of grain—he obtained similar differences in two varieties (TL 365a/25, a semi-dwarf variety, e.g.).

The tables call attention to the fact that among the varieties there are types accumulating in the grain assimilates mostly from the "leaves" and mostly from the "spike", respectively. It can also be seen that in an unfavourable crop year the role of the spike probably increases. From crosses of these types forms more productive than the parents are to be expected in theory. In experiments carried out with Swedish, Australian and Mexican varieties Derera and Stoy (1973) obtained 30–54 per cent values for the proportions of spike assimilates. The varieties in which the spike plays the greater role are characterized by a long thick awn remaining green to the end of the maturing period.

The experiment conducted by Tarchevskii et al. (1975) with a normal and a dwarf variety clearly shows the differences in photosynthetic activity between the organs (Table 32).

According to the data the dwarf variety is not inferior to the Kubanka variety with respect to organic matter production per hour, furthermore the photosynthetic intensity of spike in the dwarf variety is higher by about 50 per cent than in the tall one. The proportion of photosynthetic products carried from the flag-leaf to the spike is 27.5 per cent in the Italia 178 variety—owing to the shorter pathway of transport—and 11.3 per cent in Kubanka.

The share of the pod in the development of seed was found to be 26 per cent in bean by Crookston et al. (1974), 25 per cent in pea by Flinn et al. (1970), and 100 per cent in rape by Allen et al. (1971) (in: Derera and Stoy 1973).

Table 32

Chlorophyll (a + b) content (mg/plant) and photosynthetic rate P_N (mg CO_2 h^{-1}/plant) in spring wheat cultivars differing in stalk length (Tarchevskii et al. 1977)

Characteristics	Organ	cv. Kubanka (stem length 100 cm)		cv. Italia 178 (stem length 35 cm)	
		mg or mg h^{-1}	%	mg or mg h^{-1}	%
Chlorophyll	Ear	0.3	17	0.7	24
	Stalk	1.0	55	1.4	48
	Leaves	0.5	28	0.8	28
	Plant average	1.8	100	2.9	100
Photosynthetic rate	Ear	2.3	19.5	4.3	33.7
	Stalk	4.1	34.5	3.8	29.7
	Leaves	5.5	46.0	4.7	36.6
	Plant average	11.9	100	12.8	100

The utilization of assimilates and their distribution in the organism greatly depends on environmental factors (Maximov 1929), as suggested by the data of Table 33. These data give explanation of unfavourable changes experienced in some cases during the introduction of varieties.

From practical aspects it is decisive how long a vegetation period is required for the production of the end-product. The genetics of the *vegetation period* supplies an answer to this question when disclosing the types and course of inheritance of the property.

From the point of view of physiological genetics Samokhvalov's (1972) investigations in which early and late types are compared for physiological behaviour deserve attention (Table 34).

The data of early types unequivocally refer to a more intensive nutrient mobilization, higher photosynthetic activity and essentially lower respiration losses. Similar data on other plants have not unfortunately been encountered so far.

The correlation between productivity and length of vegetation—though most experts assume the existence of a direct relationship between them—has not been proved for all varieties. The record yields of Mexican wheat have quite recently testified that high yields can be attained in a relatively short vegetation period.

Shortening of the vegetation period (earliness) appearing as a result of crossing can be achieved in an additive way too. In some varieties the

Table 33

Effect of day-length on the distribution of assimilates in long-day and short-day plants (Maximov 1929)

Species	Day-length in hours	Weights of organs as a percentage of total weights				
		Leaves	Root	Stalk	Vegetative parts	Reproductive parts
Barley	18	12	11	55	23	77
	12	30	12	37	62	38
Millet	18	25	17	40	42	58
	12	19	10	26	29	71

Table 34

Physiological properties of seedlings in early and late varieties (Samokhvalov 1972)

Species	Variety	Organic matter requirement of 1 g seedling from the seed	Activity of chlorophyll	Intensity of respiration
Barley	Yubilenaya (early)	1.77	7.3	0.90
	Hanna (late)	1.67	5.6	1.71
Pea	56–10 (early)	2.12	8.9	2.76
	Harkovskii 131 (late)	1.89	6.2	4.86
Maize	Harkovskaya 23 (early)	1.87	16.1	1.84
	Odesskaya 10 (late)	1.71	15.1	3.84

period between sowing and flowering, while in others that from flowering to maturing is shorter. The crossing of such types is expected to result in the segregation of earlier types. These periods of development are in close connection with the heat and light periods of development, with the photoperiodic demands of varieties.

Wheat, in particular, is known to be responsive to the vernalization and light phases in respect of the length of vegetation; the latter is related to the photoperiodic demand too, though the Mexican wheat just the one that is indifferent to the duration of illumination. For the lengths of both phases

Table 35

Days of earing in F_1 hybrids of monosomic and disomic wheats (Rajki E. and Rajki S. 1969)

Variants	Chinese Spring$^+$	
	Ran. 12	Mir. 808
1A	77	83
2A	89$^+$	88
3A	74	82
4A	89$^+$	90
5A	104^{+++}	98^{+++}
6A	83	83
7A	81	86
1B	79	84
2B	81	88
3B	84	92$^+$
4B	77	84
5B	80	87
6B	84	87
7B	78	84
1D	83	84
2D	82	81$^+$
3D	77	86
4D	81	84$^+$
5D	109^{+++}	126^{+++}
6D	84	84
7D	74	82
Disome	76	86
Control	no earing	no earing
Chinese Spring	75	—

$^+$, $^{+++}$ = significant at 5 and 0.1% level, respectively.

more than one gene are responsible, that is, the vegetation period is a character of polygenic inheritance.

Morris (in: Bálint 1977) pointed out loci on every chromosome of each of the three genomes that play some role in determining the length of vegetation. In experiments conducted by Rajki E. and Rajki S. (1969) (Table 35) significant differences in the monosome series occurred in 4 cases in Rannaya 12, and 5 cases in Miranovskaya 808.

Chromosomes 5A and 5D showed the most remarkable effects. Even with as few genes as that reckoned, in the crossing populations an increase in the range of variation is expected, so to fix a new vegetation period in the progeny is no easy task.

CHAPTER 3

Nutrient utilization

Responses of varieties to nutrition

Among the tasks of plant breeding evolving varieties making the most of an *intensive* nutrient supply has come into prominence for several years.

In this context it is worth having a look at the development of rice growing in Japan (Fig. 20) and the responses of varieties to nitrogen fertilization (Fig. 21, Ishizuka 1969).

Since then the average yields have risen to about 6000 kg/ha, and in some cases yields as large as 10 ton/ha have been obtained. An increased response to fertilization is one of the elements of yield growth. The old variety Akage gave no positive yield response to increasing rates of fertilization. It is in the Fukoku and Mimasari varieties that a steady yield growth is shown. According to the data the modern varieties—owing to their more favourable genetic structure—produced larger yields even without fertilization than the old primitive ones.

The reaction to nitrogen fertilization can be easily assessed by changes in the nitrogen content. The results obtained by Nickell (1977) in experiments with sugar cane varieties are shown in Table 36.

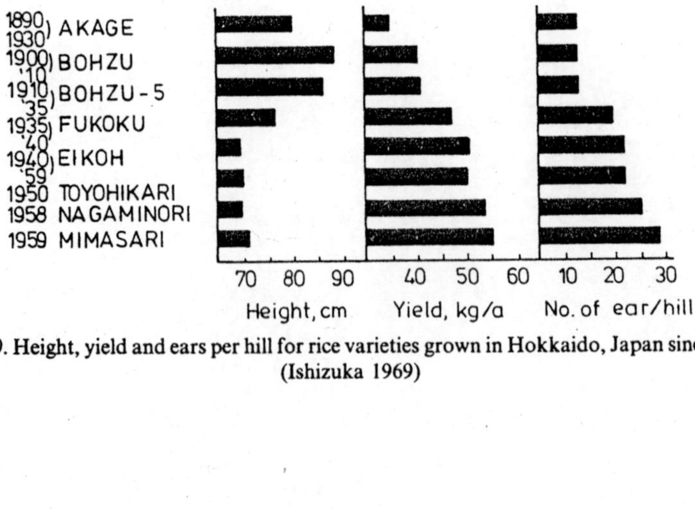

Fig. 20. Height, yield and ears per hill for rice varieties grown in Hokkaido, Japan since 1890 (Ishizuka 1969)

A comparison between the values of the first and last variety in the table reveals a linear increase of nitrogen content for the first variety and points out that the last variety does not give as high a value even with a 500 pound treatment as the other with 100 pounds. Compared to the 100 pound treatment the 500 pound treatment increased the nitrogen content of the leaf sap by 104.4 per cent in the first, and by 29.9 per cent in the last variety.

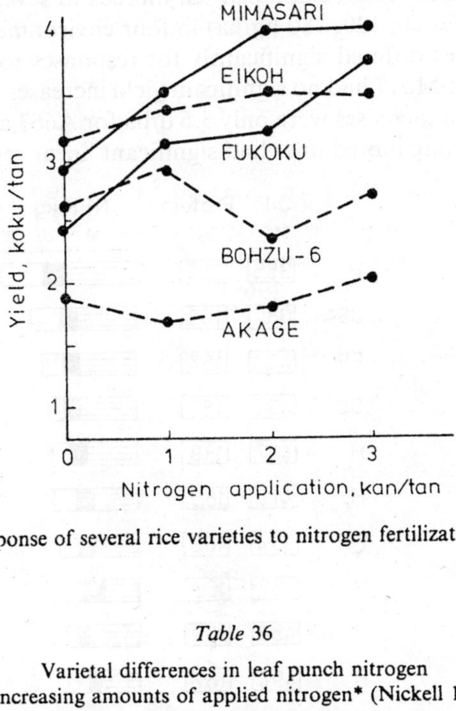

Fig. 21. Yield response of several rice varieties to nitrogen fertilization (Ishizuka 1963)

Table 36

Varietal differences in leaf punch nitrogen at increasing amounts of applied nitrogen* (Nickell 1977)

Variety	Stimulated applied nitrogen levels		
	100	300	500
56–432	0.995	1.792	2.134
49–3533	0.814	1.455	1.955
57–5174	0.946	1.438	1.899
50–7209	0.756	0.975	1.453
49–5	0.616	0.850	1.192
39–3633	0.659	0.715	0.856

* Values given are percent dry weight.

Kurnik (1978) determined the amount of NPK fertilizers required for the different sunflower varieties to produce 100 kg oil (Fig. 22).

The maximum difference between the varieties was 60 per cent, and the best hybrids produced a unit amount of oil with half as much active agent as certain open pollinating varieties. As for the seed yield the maximum difference was 50 per cent.

Russel and Balko (1980) examined ten inbreds in seven N treatments (summary of doses 60, 120, 180 kg/ha) in four environments (Table 37).

The inbred lines differed significantly for responses to N treatments. Inbreds A663 and Mo17 had no significant yield increases for N fertilizer; their average yield increases were only 3.6 q/ha for A663 and 2.4 q/ha for Mo17. All remaining inbred lines had significant linear responses. Levels

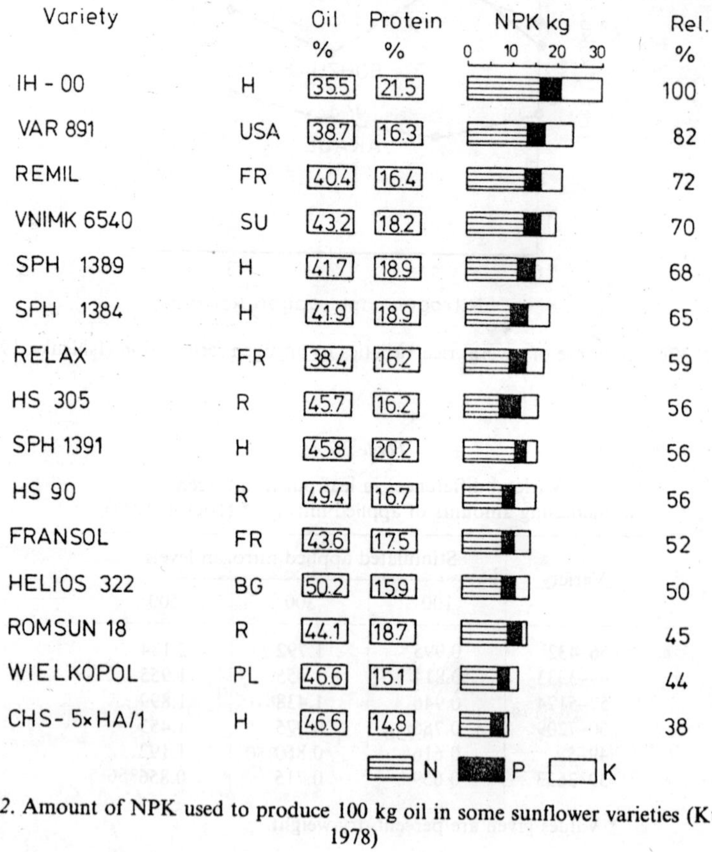

Fig. 22. Amount of NPK used to produce 100 kg oil in some sunflower varieties (Kurnik 1978)

of N fertilizer greater than 60 kg/ha caused small yield increases for A632, B70, B75, and N7A, whereas B77 had greater yield increases for higher N fertilizer levels.

B73 and Mo17 are widely used parents in the commercial production of single-cross hybrids. It is interesting, that their N response is quite different.

Rational nutrition should thus satisfy the demand of varieties rather than those of species.

Physiological effects of fertilization

The physiological effects of fertilization have the morphologically perceptible consequence of increased leaf area (LAI) and longer life of leaves (LAD).

Both types of change were demonstrated in experiments carried out with potato by Ivins and Bremner (1964, in: Evans 1975) (Fig. 23).

As a response to fertilization both LAI and LAD increased. The lower rate of fertilization seemed physiologically more economical, larger doses had minimum yield-increasing effects. Murata (1961, see Murata 1969) obtained similar results with rice.

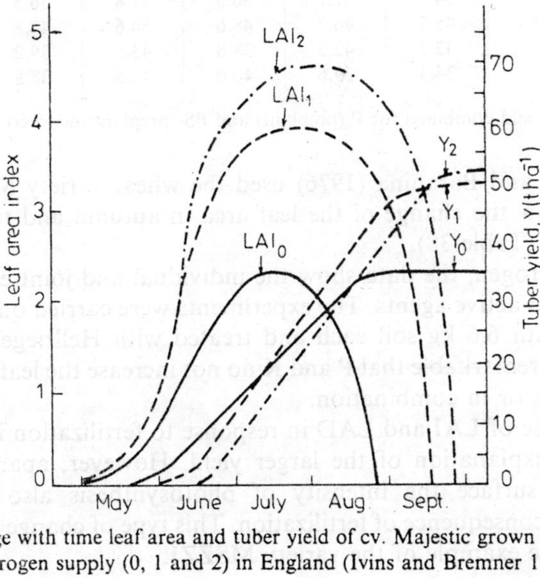

Fig. 23. Change with time leaf area and tuber yield of cv. Majestic grown at three rates of nitrogen supply (0, 1 and 2) in England (Ivins and Bremner 1964)

At the stage of milky ripeness the leaf area decreases, but with a rate of fertilization of 150 kg/ha^{-1} the proportion and absolute value of this decrease is much lower than either in the control or in the case of other fertilizer doses. In later experiments Murata and Togari (1972) pointed out a significant correlation between the yield and the leaf area index (r = +0.600 in 1967 and +0.787 in 1968). In 1969 the correlation was not significant (r = +0.260).

Table 37

Mean yields of ten inbred lines evaluated in seven N treatments, data averaged over four environments (Russel and Balko 1980)

Inbred	Nitrogen applied (kg/ha)⁺				Means
	0	60	120	180	
	(q/ha)				
A632	33.1	36.9	37.8	38.8	36.6
A663	15.4	19.2	18.6	19.3	18.1
B70	35.7	39.8	40.2	41.4	39.2
B73	44.3	52.0	52.8	51.6	50.2
B75	29.6	33.0	34.4	35.4	33.1
B76	35.3	43.6	45.0	44.6	42.1
B77	37.2	45.4	47.1	51.4	45.2
Mo17	34.7	37.1	36.5	37.8	36.5
N7A	45.7	46.7	48.6	50.6	47.8
Va26	32.2	42.2	38.8	43.8	39.2
$\bar{x}$	34.3	39.6	40.0	41.5	38.8

⁺ Yield combined for P (preplant) and PS (preplantsidedress).

Dorokhov and Baranina (1976) used the wheat variety MSZZI as a model to show the change of the leaf area in autumn and the effect of winter losses (Table 38).

Besides nitrogen, the data show the individual and joint effect of the other fertilizer active agents. The experiments were carried out in culture pots filled with 6.6 kg soil each and treated with Hellriegel's nutrient mixture. It is remarkable that P and K do not increase the leaf area either by themselves or in combination.

The increase of LAI and LAD in response to fertilization is in itself a satisfactory explanation of the larger yield. However, apart from the assimilation surface the intensity of photosynthesis also undergoes changes as a consequence of fertilization. This type of change is shown in Fig. 24 by the example of the variety MSZZI.

Table 38

Effect of fertilization on the leaf area of the winter wheat variety MSZZI in the autumn–winter period (Dorokhov and Baranina 1976)

Experimental variants	1965–1966			1966–1967		
	Autumn maximum	2. February minimum	Maximum maintained, %	Autumn maximum	30. January minimum	Maximum maintained, %
	dm²			dm²		
Control (soil without fertilizer)	0.15	0.10	67	0.39	0.11	28
2N	0.45	0.42	93	1.79	0.86	48
2P	0.12	0.06	50	0.27	0.09	33
2K	0.11	0.08	73	0.41	0.11	27
1N1P1K	0.84	0.60	71	3.08	0.53	17
2N2P	1.27	0.91	72	2.32	0.75	32
2N2K	0.65	0.29	46	1.09	0.35	32
2P2K	0.13	0.08	62	0.42	0.10	24
2N2P2K	0.90	0.56	62	3.91	0.55	14

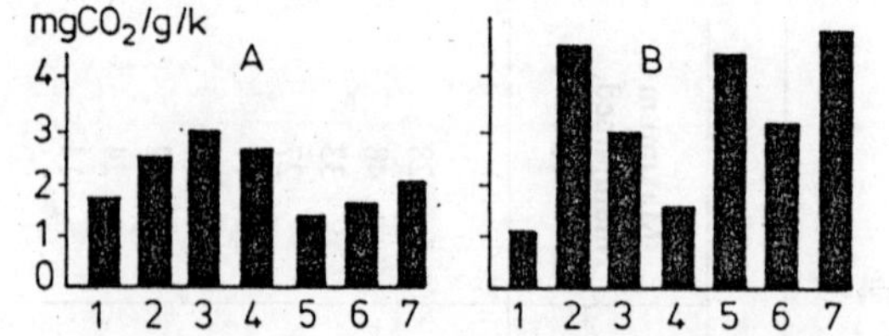

Fig. 24. Intensity of photosynthesis in spike together with the upper part of stalk in the variety MSZZI at various levels of nutrient supply; A = for 1 g dry matter; B = for 1 ear; 1 = control, 2 = +2N, 3 = +2P, 4 = +2K, 5 = +2N2P, 6 = +2N2K, 7 = 2N2P2K (Dorokhov and Baranina 1976)

It is worth comparing these columns to the crop results (Dorokhov and Baranina 1976, Table 39).

The data correspond fairly well to those in Table 38 and to certain columns in Fig. 23. The intensity of photosynthesis varied with the treatments.

Stoy (1976) carried out studies with barley. The experiments included 2 primitive barley varieties (CI 3947, CI 5195), 4 modern varieties and the local variety Nürnberg 2 (grown from seed-grains found in the foundation-stone of the Nürnberg theatre in 1832). The experiment performed in a phytotron provided a good opportunity for analysing the responses of different varieties to varying rates of N fertilization.

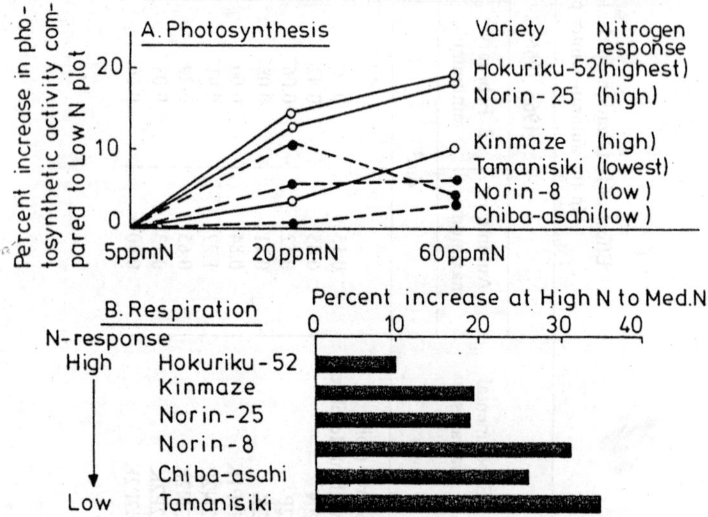

Fig. 25. Promotion of photosynthetic and respiratory activity by high nitrogen supply in rice varieties differing in nitrogen response (Osada 1966)

Table 39

Biological and grain yield (g/10 plant) and HI in winter wheat varieties given various rates of fertilization (Dorokhov and Baranina 1976)

Experimental variants	Odesskaya 3			MSZZI			MSZZI		
	1963–1964			1965–1966			1966–1967		
	Biological yield	Grain yield	HI	Biological yield	Grain yield	HI	Biological yield	Grain yield	HI
Without fertilization	4.62 ± 0.30	0.64 ± 0.15	0.14	15.46 ± 0.54	2.71 ± 0.17	0.18	6.70	1.10	0.16
+ 2N	46.85 ± 2.49	13.90 ± 1.10	0.30	85.04 ± 2.84	19.34 ± 0.89	0.23	63.70	12.00	0.19
+ 2P	4.59 ± 0.52	0.97 ± 0.18	0.21	14.69 ± 1.38	2.56 ± 0.75	0.17	7.00	1.00	0.14
+ 2K	4.59 ± 0.66	1.11 ± 0.30	0.24	16.42 ± 1.34	3.08 ± 0.21	0.19	—	—	—
+ 2N2P	28.59 ± 1.25	5.27 ± 0.15	0.18	99.48 ± 8.22	21.12 ± 1.32	0.21	31.80	6.40	0.20
+ 2N2K	44.03 ± 2.76	11.94 ± 1.17	0.27	89.49 ± 3.66	21.26 ± 0.99	0.24	50.70	12.70	0.25
+ 2P2K	3.78 ± 0.38	1.02 ± 0.21	0.27	13.21 ± 0.81	2.48 ± 0.27	0.19	5.00	0.80	0.19
+ 2N2P2K	24.06 ± 0.85	3.59 ± 0.59	0.15	84.46 ± 10.88	18.77 ± 2.72	0.22	41.20	8.90	0.21

Table 40

Effect of nitrogen (Stoy 1976)

Genotype	Nitrogen level			Quotient, High/Low
	Low	Medium	High	
	Grain yield per plant (g)			
CI 3947	0.47	0.50	0.56	1.18
CI 5195	0.38	0.48	0.51	1.34
Kristina	1.66	2.03	2.02	1.21
Bonus	1.70	1.92	2.18	1.28
Mari	1.29	1.51	1.60	1.24
Brigitta	1.25	1.49	1.79	1.43
Nürnberg 2	1.33	1.62	1.89	1.42
	Mean values of total green area per plant (cm^2)			
CI 3947	200	248	243	1.22
CI 5195	178	212	212	1.19
Kristina	217	266	276	1.27
Bonus	207	235	240	1.16
Mari	172	212	237	1.38
Brigitta	190	226	252	1.33
Nürnberg 2	202	224	274	1.37
	Mean values for specific $^{14}CO_2$-uptake ($dpm \times cm^{-2} min^{-1}$)			
CI 3947	191	212	215	1.13
CI 5195	275	266	272	0.99
Kristina	175	166	177	1.01
Bonus	149	138	143	0.96
Mari	169	152	166	0.98
Brigitta	175	168	170	0.97
Nürnberg 2	179	174	179	1.00

The results are summed up by Stoy (1976) in Table 40.

Surprisingly enough, it is only in the variety CI 3947 that any considerable difference was found in the intensity of CO_2 uptake between low and high N-rate treatments; yield differences may rather be attributed to changes in the leaf area. In barley—like in some wheat varieties—the intensity of photosynthesis is higher in the primitive than in the cultivated varieties.

Osada (1966) calls attention to the fact that varieties making good use of nitrogen are characterized not only by increased intensity of photosynthesis, but by a moderate increase in the respiration losses too (Fig. 25).

In this experiment with rice, on increasing rates of nitrogen fertilization an almost linear rise in photosynthetic activity was observed in the varieties. There was a variety in the experiment (Norin-8) which considerably increased its photosynthetic activity in response to a moderate rate of N application and reduced it when the doses grew further. From the point of view of the final result, however, the respiration loss is of decisive importance. In the variety Hokoruku-62 the increasing photosynthetic activity is accompanied by a moderate increase of respiration loss. The worst varieties, on the other hand, are characterized not only by a low photosynthetic activity shown in response to increasing rates of N fertilization but also by a substantial increase in the respiration loss.

It is thus clear that from the point of view of nitrogen utilization those varieties are favourable in which the increasing nitrogen doses induce a gradual increase in the photosynthetic activity, and the parallel increase of respiration loss is of much lower rate.

Apart from increased organic matter production it is important in what part of the plant the nitrogen will be incorporated. Yamada (1959) studied this problem with rice. Of the varieties included in the experiment Norin 1, Taichung-65 and Kanan-8 were of japonica while N-302 and H-4 of indica type (Fig. 26).

The daily rate of dry matter production showed a substantial increase in each variety in the case of large N doses. Lower rates of N application increased the dry matter production only in the two indica type varieties. The incorporation in the grain was less satisfactory. With the exception of Norin-1 and H-1 larger doses involved lower nitrogen utilization. The positive effect of the lower N rate was only felt in two cases (Taichung-65 and Kanan-8).

On the basis of the analysis of data the breeder would cross the varieties Norin-1 and M-302 in order to obtain a more favourable combination with respect to the parameters examined.

From the data presented one can clearly see that all the trends of LAI, LAD, photosynthesis and respiration intensity ought to be known to understand the N-responses of the different varieties. Such experimental data are, unfortunately, hardly available.

From the point of view of breeding it is also useful to know whether the individual lines improve or reduce the extent of N utilization in the combinations produced. Pollmer et al. (1979) studied the effect of 10 smooth-grained (flint) and 9 dent lines on the GCA general combining ability. Compared to the average 194.2 the GCA ranged between +27 and −21.4 in response to flint and from +45.6 to −26.7 under the influence of dent lines..

Besides the above, the factors that influence the nitrogen uptake ought to be known exactly. All of them cannot be dealt with here, only the role of certain *enzymes*—of the N nitrate reductase (NR) in the first place—will be discussed.

Trying to find an explanation of differences between the maize varieties in the extent of nitrogen utilization Hoener and De Turk (1938, in: Vose 1963) supposed that the different responses of various maize varieties to increasing rates of N application were due to the fact that plants with high protein content had a larger nitrate-reducing enzyme capacity than the

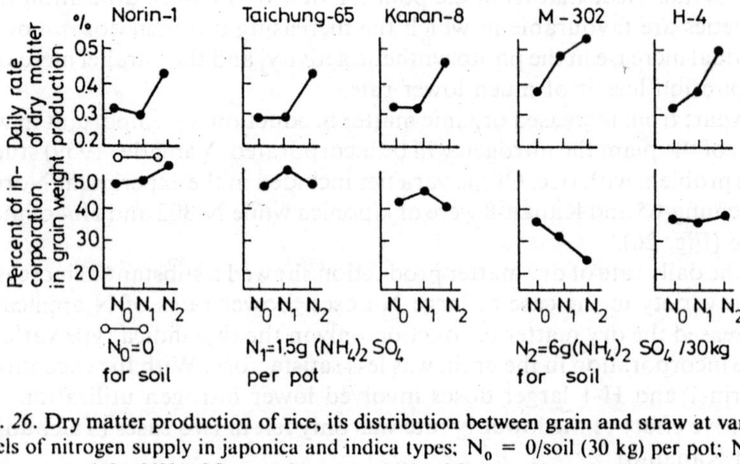

Fig. 26. Dry matter production of rice, its distribution between grain and straw at various levels of nitrogen supply in japonica and indica types; N_0 = 0/soil (30 kg) per pot; N_1 = 1.5 g$(NH_4)_2SO_4$/pot; N_2 = 6 g$(NH_4)_2SO_4$/pot (Yamada 1959)

"low protein" plants. In the case of lower N-rates a smaller amount of nitrate reductase may be sufficient to reduce the nitrate, the difference appears therefore only with higher rates of nitrate fertilizer. Vose and Breese (in: Vose 1963) found higher nitrate concentrations in brome-grass varieties giving poor responses to nitrogen fertilization than in the highly reactive types which suggests the varying activity of the nitrate reductase enzyme.

Hageman and Gitter (in: Hageman et al. 1974) pointed out a consequently higher nitrate reductase activity in the Hy2 × OH7 hybrid

Table 41

Trends of grain yield and protein content in six maize hybrids grown at two different stand densities in the case of nitrogen** top dressing (Deckard et al. 1973)

Number of hybrids	Time of N application (number of days after sowing)	59,000 plants/ha		70,000 plants/ha	
		Grain yield* (kg/ha)	Protein (%)	Grain yield* (kg/ha)	Protein (%)
4	control	10,331	7.6	10,048	7.7
4	44	10,645	9.4	10,500	9.6
4	59	10,400	9.4	10,029	9.2
4	72	10,852	8.9	10,155	9.3
1	control	10,161	7.7	10,218	7.8
1	44	10,780	9.0	10,236	9.1
1	59	10,243	9.1	10,425	9.1
1	72	10,500	9.3	10,638	9.0
2	control	9,225	8.8	9,169	8.2
2	44	9,577	10.3	9,307	10.6
2	59	9,684	10.4	8,692	11.1
2	72	9,483	10.2	8,924	10.1
5	control	8,886	8.9	8,566	8.4
5	44	9,477	11.1	9,640	11.4
5	59	10,192	11.2	6,685	11.5
5	72	9,068	11.3	8,817	10.8
6	control	8,666	9.1	8,698	9.1
6	44	9,244	10.8	8,610	10.9
6	59	9,093	11.3	8,095	10.9
6	72	9,012	10.8	8,057	11.4
3	control	9,458	7.9	7,756	7.8
3	44	8,604	8.1	7,611	8.4
3	59	9,018	8.1	8,013	9.1
3	72	8,836	8.4	8,114	8.6

LSD (at 0.05% level) (a) on N top dressing between hybrids and stand densities 800 kg/ha for grain yield and 0.63% for protein; (b) between the mean values of hybrids 338 kg/ha for grain yield and 0.35% for protein.

* = with 15.5% water content.

** = 337 kg N/ha in the form of nitrate.

compared to WF 9 × C103. The higher nitrate reductase activity showed a positive correlation with the water soluble proteins, and a negative one with the nitrate content.

Data on the total yields and protein contents of six hybrids 44, 59 and 72 days after sowing are seen in Table 41 and the correlation between grain yield and nitrate reductase activity in Fig. 27 (Deckard et al. 1973).

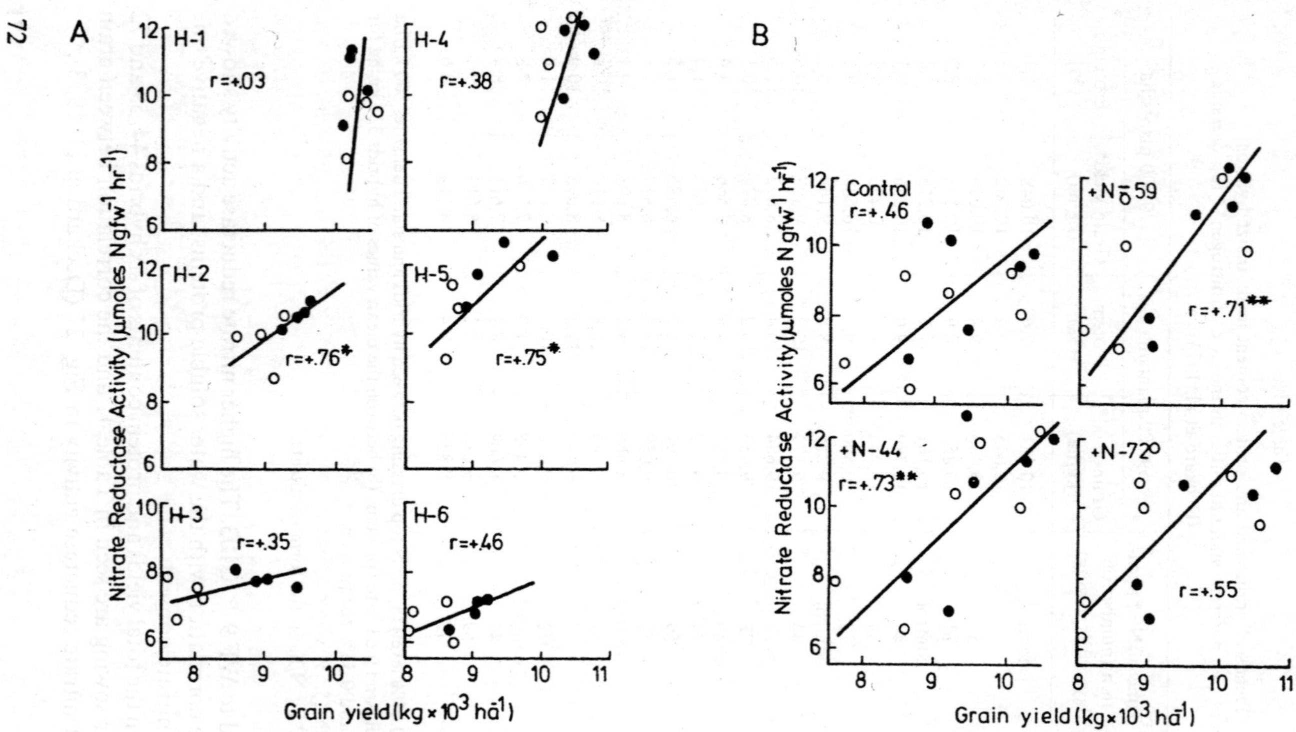

Fig. 27. **Relation between seasonal average nitrate reductase activities and grain yields during vegetation; A = grouped by hybrids; B = grouped by nitrogen treatments. Code for populations ● and ○ for 59,300 and 79,000 plants ha^{-1}, respectively (Deckard et al. 1973)**

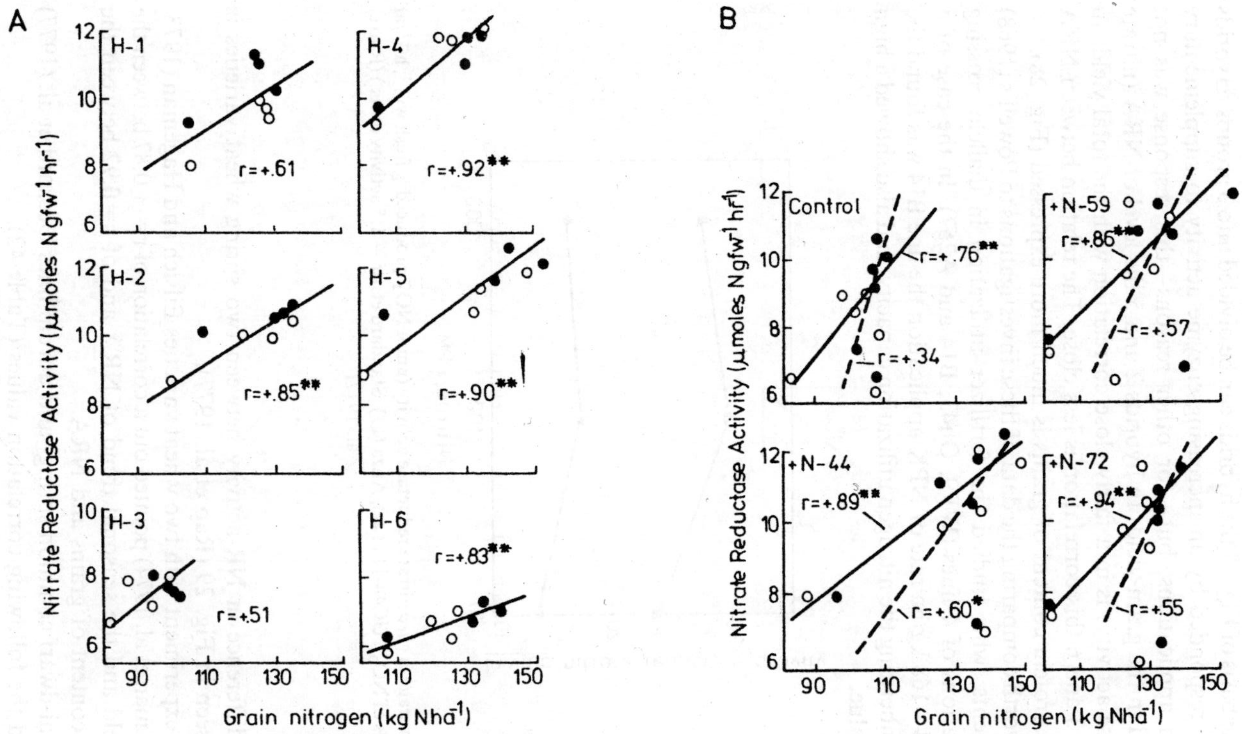

Fig. 28. Relation between protein content of grain and average nitrate reductase activities during vegetation; A = grouped by hybrids; B = grouped by nitrogen treatments. Code for populations ● and ○ for 59,300 and 79,000 plants ha^{-1}, respectively (Deckard et al. 1973)

On the basis of Fig. 27 the hybrids can be divided into 3 groups. Hybrids 1 and 4 responded by an increasing enzyme activity to supplementary nitrogen applications, but—for other reasons—this response was not reflected in the grain yield. In hybrids 2 and 5 the level of NRA (nitrate reductase activity) is in a fairly close correlation with the total yield; in hybrids 3 and 6 this correlation is less close. The relation between NRA and the protein content of grains is much more expressed (Fig. 28).

It is worth comparing the data of these investigations to Powell's (1968) experiments, who studied the fertilizer utilization in diallele crossing combinations of 4 lines (WF9, Oh43, B14 and A257). In the case of a 224—84—168 kg/ha rate of NPK application the line B14 was found the best in inheriting the nitrogen utilization capacity, and also showed a high NRA value.

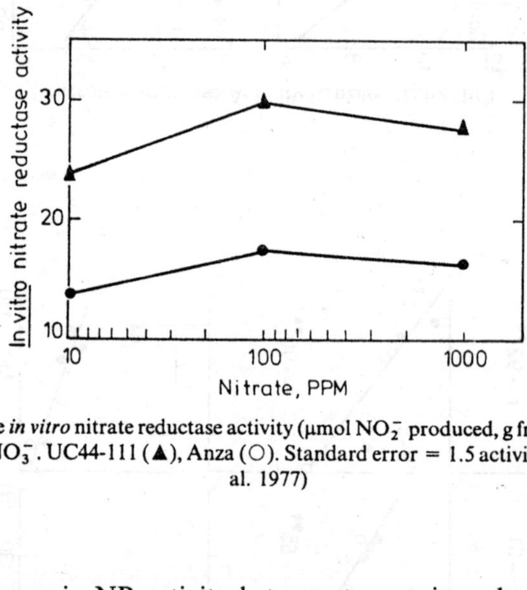

Fig. 29. Average *in vitro* nitrate reductase activity (μmol NO_2^- produced, g fresh wt^{-1}, h^{-1}/at three levels of NO_3^-. UC44-111 (▲), Anza (○). Standard error = 1.5 activity units.) (Rao et al. 1977)

The difference in NR-activity between two spring wheat varieties is clearly seen in Fig. 29 (Rao et al. 1977).

In an experiment with two wheat varieties Eilsich and Hageman (1973, in: Hageman et al. 1974) pointed out a correlation of r= +0.87 between the total yield and the seasonal trend of NRA and of r=0.92 between the protein content of grains and NRA.

In semi-dwarf common spring- and durum lines Deckard et al. (1977) obtained the following correlation values (Table 42).

His data correspond to the opinion of Rao et al. (1977) for the NR-activity by itself is not a sufficient criterion for the breeder when selecting a segregating population for higher productivity.

The absence of a direct relationship can also be explained by the fact that the value of NRA depends on the amount of molybdenum as an electron carrier. So the NR-activity in a variety is primarily determined by the amount of Mo taken up by the plants.

Experiments with barley, on the other hand, suggest an antagonism between the photosynthesizing enzymes and the nitrate reductases (Fig. 30, Warner and Kleinhofs 1974). These ratios vary from variety to variety.

Apart from these contradictory data, increasing the NR-activity may be advantageous in breeding plants for improved physiological genetic features.

Table 42

Correlation coefficients between nitrate reductase activities and grain yield, percent grain protein, and grain protein yield (Deckard et al. 1977)

Variables correlated	HRS lines	Durum lines	
		All	Without 1 pair
	values (r)		
NRA (seasonal average) and:			
Grain yield	0.457**	0.278	0.585**
Percent grain protein	−0.562**	−0.008	−0.532**
Grain protein yield	0.320	0.358	0.397
NRA (28 days after seeding) and:			
Grain yield	0.524**	0.314	0.693**
Percent grain protein	−0.510**	0.010	−0.470*
Grain protein yield	0.425*	0.410*	0.592**
No. of observations	32	30	24

*,** Significant at the 0.05 and 0.01 levels, respectively.

In breeding experiments little attention has been paid so far to the fact that genetic changes (e.g. in induced mutants) may involve changes in the nitrate reductase activity (NRA) too. Tokarev and Shumnyi in 1977 produced mutants in the barley variety Viner by using ethyl-methane-sulphonate. The experiments were aimed at finding plants in which the NRA was low or completely absent. This demand was met by 71 of a total

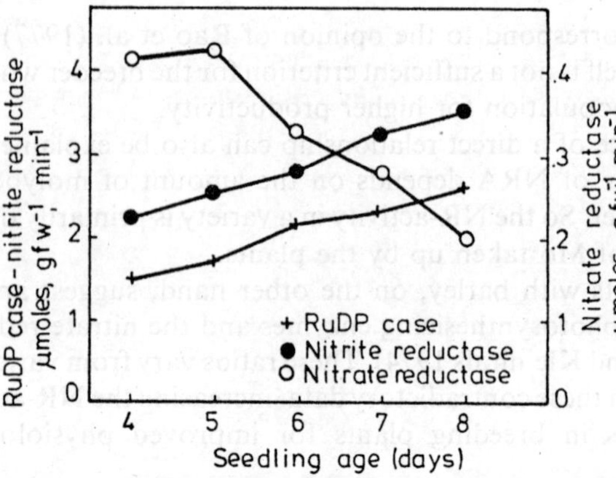

Fig. 30. Effect of seedling age of seedling on nitrate reductase, nitrite reductase, and ribulose diphosphate carboxylase activities in Himalaya barley (Warner and Kleinhofs 1974)

of 5450 M_2 plants. On the basis of analysing the leaves of 10 day-old seedlings from 50 of the 71 plants, progenies producing grain could be raised. In families selected from their M_3 generations the level of NRA showed the following trend (Table 43).

Table 43

Nitrate reductase activity in M_3 families (micro-M NO_2/g h) (Tokarev and Shumnyi 1977)

M_3 families	Variation of NRA by family ($NR_{min.}$—$NR_{max.}$)
LN 1	0.4–1.4
LN 4	0.5–2.4
LN 17	0.2–1.4
LN 29	0.2–2.5
LN 57	0.1–0.3
Control	3.8–4.6

According to the evidence of the data the NR-activity could be substantially reduced compared to the control. Unfortunately, there is no reference to how this change affected the volume of yield, and the authors did not make selection for mutants with a higher NR-activity than that of the control.

Few physiological genetic studies are known to have been carried out on the utilization of phosphorus. Table 38 and Fig. 24 include some data on P, mainly to show that favourable effects are mostly expected from the joint application of nitrogen and phosphorus.

Nelson (1966) carried out investigations in Florida on the N, P and K utilization of a number of maize hybrids (Table 44).

Table 44

Uptake and recovery of N, P and K by 5 corn hybrids (Nelson 1966)

	DeKalb 805	Pioneer 309 B	Dixie 18	Florida 200	Coker 67
Nitrogen (300 lbs. N/Acre)					
Uptake, lbs./Acre	116.0	141.0	134.0	176.0	166.0
Percent recovery	30.7	35.5	34.1	44.2	42.1
Phosphorus (200 lbs. P_2O_5/Acre)					
Uptake, lbs./Acre	32.1	36.6	43.5	61.8	50.4
Percent recovery	10.3	11.4	12.6	20.1	14.7
Potassium (300 lbs. K_2O/Acre)					
Uptake, lbs./Acre	157.2	208.8	208.8	204.0	224.4
Percent recovery	42.6	55.6	53.6	51.6	60.3
Forage yield/Acre					
Dry weight × 1000	12.5	12.2	15.0	15.4	14.3

From the data two conclusions can be drawn: the largest yield and the highest N and P-utilization were found in the variety Florida 200; as for the utilization of potassium this variety was placed in another group. There is no absolutely direct relationship between high nutrient utilization and large yield; the Dixie 18 hybrid for example while inferior to the hybrid Coker 67 in nitrogen and phosphorus utilization exceeds it in productivity.

There was a maximum of about 50 per cent difference in potassium utilization between the maize varieties (Table 44). According to Fig. 24 the intensity of photosynthesis related to fresh weight was increased essentially and that calculated for the ear moderately by potassium.

The microelements are important factors of an adequately balanced nutrient supply. Particularly with higher rates of fertilizer application it was found that microelements reach a minimum, therefore the demands of plant species and varieties must be known in this respect, too.

Differences in the uptake and utilization of microelements between various maize lines are shown in Table 45 compiled by Clark (1973). The efficiency is related to the microelement demand of the line. In Clark's (1973) experiments a line required 50 ppm while a free-pollinated variety only 10 ppm concentration for normal growth.

Weiss (in: Vose 1963) examining Manchurian soya varieties grown in calcareous soil found great differences in chlorotic spots. Under experimental conditions these differences appeared when the degree of Fe utilization was low. According to the results of experiments with F_2-, F_3- and back-crossed generations the mentioned characteristic was determined by a single gene. On crossing types of high and low Fe-utilization the author observed total dominance in the F_1 generation. In plants making little use of iron, higher pH-values and smaller quantities of Fe were found than in those responsive to iron.

In the course of investigations aimed at determining the magnesium demand genotypic differences were found in the apple varieties.

Table 45

Relative efficiencies of corn inbred lines for uptake and utilization of mineral elements (Clark 1973)

Element	Inbred group	Most efficient	Least efficient
Fe	I	WF9, MS1334, Pa32	ys_1/ys_1,Pa54, IaB8
	II	A554, A619	A632, A635
	III	W153R, A239	K55, CI66, K64
Zn	I	Oh40B, WH	IaB8, Pa32
	II	H84, A619, N28, A554	A632, B14A, W64A, A635
	III	W153R, B57	K55, C103
Mg	I	CMD5, Pa32, ys_1/ys_1	Ind38–11, Oh40B, Pa94, IaB8
	II	N28, H84, B37	A635, W64A, B14A
	III	B57, CI64, A239	C103, C123, A251
Ca	I	WF9, WH	Pa36, Pa32
	II	B37, Oh43	N28, A554
	III	B57, K64	A251, C123, W117
		Most tolerant	Least tolerant
Al (toxicity)	I	PaB8A	Pa54, PaW703
	II	B37, Oh43	A635,A554
	III	W117, Mo17	C103
		High accumulation	Low accumulation
Mn	I	Pa54, PaW703	Pa83
	II	B14A, B14	W64A, N28, H84
	III	B57, W153R	CI64

Magnesium reactivity in celery (Pope et al. 1953 in: Vose 1963) is determined by a single gene. In the Utah 10B variety this gene is recessive.

Sleper et al. (1977) studying the magnesium deficiency disease of tall fescue (*Festuca arundinacea* Schreb.) followed the transmission of Mg,

Table 46

Mean concentrations of minerals in the forage of 11 fescue clonal lines used as parents in DI and DII diallels (Sleper et al. 1977)

Mineral+	Harvest	DI parents			DII parents		
		1974	1975	Mean	1974	1975	Mean
Mg	First	0.163	0.166	0.165	0.165	0.178	0.172
	Second	0.175	0.236	0.206	0.192	0.251	0.222
	Third	0.197	0.236	0.217	0.197	0.263	0.230
Mean		0.178	0.213	0.196	0.185	0.231	0.208
Ca	First	0.263	0.314	0.289	0.272	0.299	0.286
	Second	0.396	0.496	0.446	0.411	0.533	0.472
	Third	0.442	0.367	0.405	0.463	0.403	0.433
Mean		0.367	0.392	0.380	0.382	0.412	0.397
K	First	1.871	2.311	2.091	1.914	2.290	2.102
	Second	1.486	1.660	1.573	1.549	1.621	1.585
	Third	1.282	1.553	1.418	1.354	1.888	1.621
Mean		1.546	1.841	1.694	1.606	1.933	1.769
P	First	0.312	0.336	0.324	0.307	0.326	0.317
	Second	0.328	0.384	0.356	0.329	0.365	0.347
	Third	0.247	0.279	0.263	0.241	0.291	0.266
Mean		0.296	0.330	0.314	0.292	0.327	0.310
K/(Ca + Mg)	First	1.819	2.035	1.927	1.827	2.022	1.925
	Second	1.133	0.977	1.055	1.107	0.908	1.008
	Third	0.879	1.067	0.973	0.913	1.178	1.046
Mean		1.277	1.360	1.318	1.282	1.368	1.326

+ Concentrations: Mg, Ca, K and P (%) and K/(Ca + Mg) (mg/100 g dry matter).

some macroelements and the ratio of K(Ca + Mg) in 6 clones each. It is worth looking at the concentrations of the individual elements at three dates of cutting (13 May, 2 July, 23 October 1974: 13 May, 8 July and 5 November 1975), and at the heritability values (Tables 46 and 47).

Since the K(Ca + Mg) ratio is markedly a hereditary feature it is possible to produce a variety that remains healthy even at a low level of magnesium.

Table 47

Heritability estimates calculated by progeny-parent regression (1) and the standard unit (2) methods (Sleper et al. 1977)

Mineral	Year	Heritability (%)					
		First harvest					
		(1)	(2)	(1)	(2)	(1)	(2)
DI							
Mg	1974	82	68	86	92	85	94
	1975	103	74	113	93	113	89
Ca	1974	0	0	112	78	75	53
	1975	37	35	122	87	91	76
K	1974	37	38	69	58	52	38
	1975	28	29	76	80	110	32
P	1974	35	39	99	73	82	45
	1975	34	33	75	56	110	73
K/(Ca+Mg)	1974	60	44	88	88	90	84
	1975	70	75	86	92	89	57
DII							
Mg	1974	57	50	74	83	69	84
	1975	52	49	57	26	65	60
Ca	1974	84	71	80	64	69	64
	1975	75	73	91	90	43	44
K	1974	91	78	46	60	78	78
	1975	86	61	62	70	54	61
P	1974	74	78	67	74	134	84
	1975	54	69	78	80	167	63
K/(Ca+Mg)	1974	97	83	84	81	65	86
	1975	69	75	31	41	45	61

As regards the demand for boron, varietal differences were found in grape-vine, sugar-beet and celery. In celery, susceptibility to boron deficiency–just like in Mg—is determined by a single gene (Vose 1963).

As seen from the data presented, there are substantial differences between the varieties in the utilization of nutritive elements. With respect to the nature of inheritance very little has been published. Owing to the large number of determinative factors it is most probably a case of polymer inheritance.

CHAPTER 4

Water utilization

Our farm crops use carbon dioxide present in the air, water, macro- and microelements to build up their organisms with the help of solar energy. An average of 1.5 g CO_2 is used to produce 1 g of dry matter. This amount of CO_2 is obtained from 2.5 m^3 air. With higher CO_2 content photosynthesis can be intensified (Pál 1973). We have also seen the role nutritive elements play in this process.

The third resource is water. Let us take a less known role of water. Hoffmann (1975) established that 48 per cent of the solar energy reaching the leaves serves the water economy of leaves.

Water absorbed by the roots leaves the plant through the stomates (except the part incorporated in the organism). The stomates being openings for the gas exchange establish the connection of the two processes. The variability of photosynthesizing ability is already known. With respect to water utilization we have to know that the amount of dry matter formed after the evaporation of 1 litre of water ranges from 1 to 8 g (productivity of transpiration) depending on the species. A more frequently used index is the transpiration coefficient, the reciprocal of the former value, which expresses how many ml of water are consumed while 1 g of dry matter is formed.

These values show a close relationship with the number and size of stomates (Table 48, Pál 1973).

Three aspects of the genetic differences of water utilization will be dealt with here: economy of water utilization (transpiration coefficient), response to irrigation, and response to water deficiency.

It is well known that there are essential differences in water utilization in the plant species.

Ringoet (1952, in: Alvim and Kozlowski 1977) determined transpiration coefficients for several plant species in Yagambi (Zaire) which are shown in Table 49.

Table 48

Number and dimensions of stomata in various plants
(Pál 1973)

Plant	Stomata				Area of stomata (μ^2)	Total area of stomata as a percentage of the leaf surface
	(Number/mm²)		Length (μ)	Width (μ)		
	Upper	Lower				
	Leaf surface					
Oat	25	23	38	8	239	0.98
Wheat	60	41	38	7	209	0.52
Maize	52	68	19	5	75	0.88
Sunflower	58	156	22	8	136	3.13
Tomato	12	130	13	6	61	0.85
Bean	40	281	7	3	17	0.54
Apple	—	400	14	12	132	5.28

Table 49

Mean transpiration coefficients of various plants at Yagambi
(Ringoet 1952, in: Alvim and Kozlowski 1977)

	Mean transpiration coefficient	Range of mean for 11–19% soil moisture
Oil palm (*Elaeis guineensis*)	294	115
Upland rice (*Oryza sativa*)	413	50
Robusta coffee (*Coffea canephora*)	578	20
Palisotha thyrsiflora (wild plant)	618	10
Cacao (*Theobroma cacao*)	866	—

As seen from the data the oil-palm possesses two advantageous properties: it has a low transpiration coefficient and can be well accommodated to changing values of soil moisture.

In experiments carried out with vegetables Cselőtei (in: Somos 1966) obtained average values of 262 for bean and 331 for paprika. The values published by Szalai (1974) are 198 for cabbage and 282 for carrot. We can only change an unfavourable situation by finding types better utilizing the water within the species. As an example we refer to the results obtained by Somos (1966) in experiments with paprika (Table 50).

These values vary with the soil type. On quartz sand, sandy loam and loam the Szegedi red pepper variety gave values of 270, 318 and 421, the tomato-shaped green paprika 318, 322 and 258, respectively. In Cselőtei's (1964) experiments an average value of several years of 354 was obtained for the paprika variety Cecei édes.

Table 50

Trend of the transpiration coefficient of various types of paprika grown on quartz sand (Somos 1966)

Name of variety	Time of measuring			Average
	27. July	26. August	23. September	
Szegedi fűszer	270	255	285	270
Dokomlási	339	307	284	310
Cecei édes	336	311	342	330
Kalinkói zöld	377	323	329	342
Paradicsom alakú zöld	327	317	309	318
Tétényi	350	406	396	384
Csokros zöld	369	390	355	371
Elefántormány	357	350	311	339

The effects of various rates of potassium and nitrogen application on productivity and transpiration coefficient and the interaction of the two factors in 6 winter wheat varieties included in culture pot experiments conducted by Zwicker (1976) are shown in Fig. 31.

According to the data there was a highly significant negative correlation ($r = -0.806^{+++}$) between productivity and transpiration coefficient. The largest yield was obtained with the varieties Winnetou and Fakir, the smallest one with Poros and Pluto.

In an experiment carried out with 3 cucumber varieties (K. high-yielding, Hokus, Delicatess) Cselőtei (1978) observed a uniform water utilization and determined similar transpiration coefficients, while as for the volume of yield the varieties showed the following order: K. high-yielding exceeded Hokus by about 10, and Delicatess by some 30 per cent. There were differences between the varieties in the amount of organic matter incorporated in the crop (HI).

The effects of rainfall and nitrogen supply on grass species and varieties were studied by Burton et al. (1957). The results are contained in Table 51.

An increase in the nitrogen supply resulted in a much more efficient utilization of water (except the Pangola grass). The two most drought

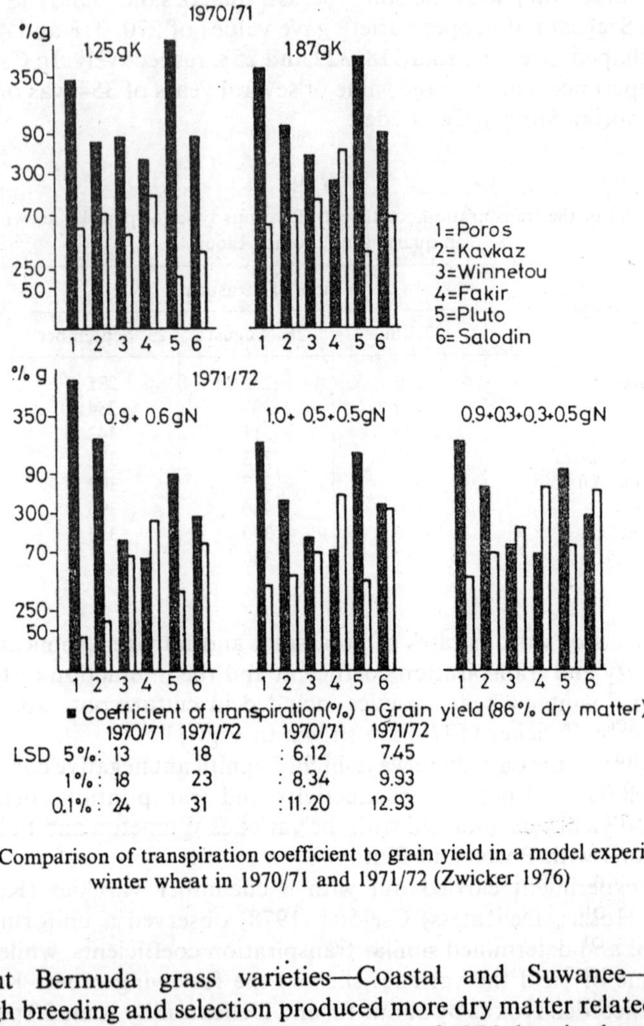

Fig. 31. Comparison of transpiration coefficient to grain yield in a model experiment with winter wheat in 1970/71 and 1971/72 (Zwicker 1976)

tolerant Bermuda grass varieties—Coastal and Suwanee—evolved through breeding and selection produced more dry matter related to unit amount of water evaporated in the dry year of 1954 than in the wet 1953. The common Bermuda grass, on the other hand, exhibited a less efficient water utilization under dry conditions than when the precipitation conditions were favourable. It is worth noting the data of the table which indicate how much high rates of nitrogen application lowered the transpiration coefficient improving thereby the water utilization of the plant.

Table 51

Effect of different grasses, rainfall and nitrogen levels on water use efficiency (Burton et al. 1957)

	(kg water/kg dry matter)					
	N levels (kg/ha) 1953 (1007 mm rain)			N levels (kg/ha) 1954 (347 mm rain)		
	56	112	224	56	112	224
Cynodon dactylon Common	6,812	2,896	1,546	9,738	5,028	4,336
Cynodon dactylon Coastal	2,478	1,431	803	1,547	982	641
Cynodon dactylon Suwannee	1,923	1,104	692	1,107	725	452
Paspalum notatum Pensacola	2,200	1,321	870	3,103	1,945	1,239
Digitaria decumbens Pangola	2,249	1,585	2,240	2,843	2,513	3,016

Effect of irrigation

In many countries of the world natural precipitation must be supplemented by irrigation. It is, however, not negligible how much the cultivated varieties are responsive to irrigation.

We have shown already in Fig. 11 how irrigation increases the leaf area and the life-time of leaves. The experiment performed by Petinov and Korshunava (in: Petinov 1965) with maize showed how the productivity of photosynthesis became higher with an increase in the leaf area (Fig. 32). In consequence of this—as pointed out by Kurnik (1970) with pea (Table 52) and by Koltay and Balla (1975) with wheat—the yield generally increases, though not in all varieties (Table 53).

In experiments with wheat, the differences between years and varieties, respectively, mostly ranged from 5 to 20 per cent. Kavkaz, Avrora and Libellula were found to be the most responsive to irrigation both in 1972 and 1973. Other wheat varieties (Martonvásár 1, Triticale) gave opposite responses in the two crop-years. The difference in responsiveness to irrigation which can be traced to more than one factor, is totally incidental since neither the latter varieties nor the highly responsive ones were selected for irrigation.

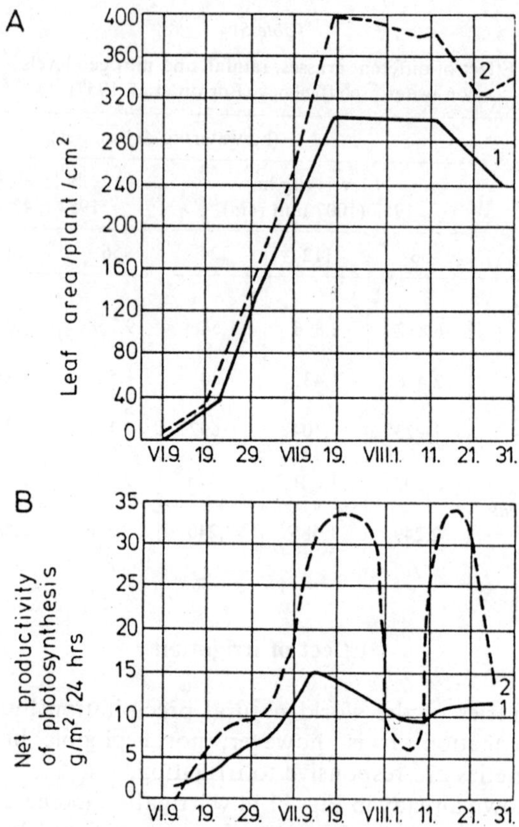

Fig. 32. Effect of irrigation on the leaf area (cm²/plant) (A) and on the net productivity of photosynthesis in maize leaves (g/m²/24 krs (B); 1–nonirrigated; 2–irrigated (Petinov 1965)

Table 52

Effect of irrigation on yield in pea varieties in Iregszemcse (Kurnik 1970)

Variety	Dry		Irrigated	
	(q/ha)	Ratio	(q/ha)	Ratio
Korai Viktória	17.3	100	2.13	123.0
Kurz und Gut	13.5	100	16.4	122.0
Serpette	19.4	100	23.7	122.0
Express	11.6	100	13.1	120.8
IP 2	17.5	100	20.4	117.0
Iregi yellow fodder	19.5	100	21.3	108.9
Pauli	17.8	100	16.9	95.1
Rondó	17.6	100	16.6	94.1

Table 53

Grain yield response of wheat varieties to irrigation in Martonvásár (Koltay and Balla 1975)

Variety	Grain yield on the average of 32 series		D (kg/ha)
	Non-irrigated	Irrigated	
	(kg/ha)		
1972			
Avrora	5,526	6,006	480
Kavkaz	5,319	5,999	680
Yubilenaya 50	4,879	5,213	334
Mironovskaya 808	4,439	4,786	347
Martonvásári 3	4,226	4,433	307
Libellula	4,159	4,613	454
Martonvásári 2	4,039	4,286	247
Martonvásári 1	4,006	4,033	−27
Fertődi 293	4,266	4,206	−60
Triticale	3,786	4,333	547
1973			
Kavkaz	6,206	7,446	1,240
Yubilenaya 50	6,193	7,346	1,153
Mironovskaya 808	5,999	6,806	807
Martonvásári 1	5,853	7,166	1,313
Avrora	5,833	7,293	1,460
Libellula	5,733	6,913	1,180
Martonvásári 2	5,606	6,693	1,087
Fertődi 293	5,473	6,293	820
Martonvásári 3	5,353	6,646	1,293
Triticale	4,146	4,479	333

Effect of water deficit

The photosynthetic processes are essentially influenced by the water deficit. The extent of water deficiency resulting in the closing of stomates and limiting the photosynthesis is seen in Table 54 based on the data of Strebeyko (1973).

Reduction in photosynthesis appears in maize beyond an 8 per cent water loss, and is accompanied by an increase in the intensity of respiration (Fig. 33).

Similarly to the species there are differences between the varieties, as shown by the results of experiments carried out by Sammons et al. (1978) with soya plants.

Table 54

Water deficits limiting the photosynthesis (Strebeyko 1973)

Plant species	Percentage value of water deficit
Cabbage	20
Rape	25
Nicotiana rustica L.	30
Sugar beet	35–40

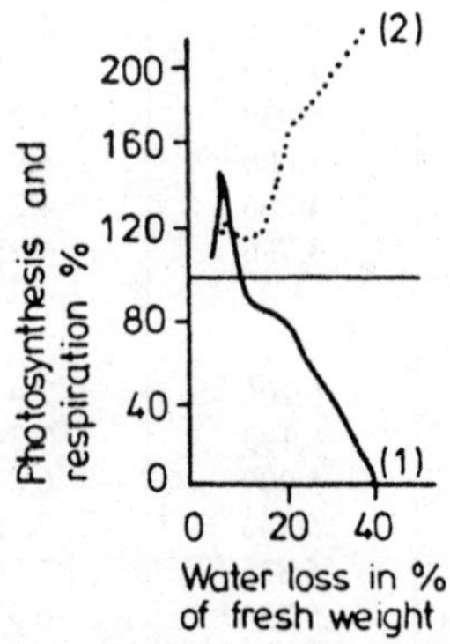

Fig. 33. Effect of water deficit on the intensity of photosynthesis in maize leaves; 1–photosynthesis; 2–respiration (Andrejenko and Kuperman 1961)

In a pot experiment, the authors studied the responses of 21—22 day-old plants to various extents of water deficiency. The effect of water deficiency on the surface area (LSA) and dry weight (LDW) of leaf is seen in Table 55.

Differences in drought tolerance between the varieties are equally well expressed by the two indices (the extent of decrease in them in response to water deficiency is not the same).

The trend of yield is also influenced by the extent to which water deficiency decreases the intensity of photosynthesis. This correlation is shown in Table 56.

Table 55

Mean leaf lamina surface area (LSA) and dry weight (LDW) for individual cultivars measured at three soil water potentials. Means are for sums from six plant replicates (Sammons et al. 1978)

Cultivars	Soil water potential in bars					
	−0.7	−3.0	−8.9	−0.7	−3.0	−8.9
	(dm^2m^2)			(g)		
MG 0–00						
Cayuga	2.76a*	1.98b	1.90b	1.04a+	0.76b	0.73b
Grant	4.19a	3.28b	3.14b	1.40a	0.13b	1.09b
Manitoba Brown	4.42a	3.66b	3.46b	1.54a	1.20b	1.12b
Pagoda	3.54a	2.78b	2.32c	1.08a	0.94b	0.76c
MG I						
A 100	2.88a	2.48b	2.04c	1.30a	1.17b	0.96c
Black Eyebrown	2.36a	2.12a	1.60b	0.86a	0.88a	0.72b
Blackhawk	3.03a	2.36b	2.02c	1.18a	0.97b	0.80c
Chippewa 64	2.92a	2.34b	2.40b	1.16a	0.88b	0.84b
Disoy	4.36a	3.75b	2.93c	1.75a	1.54b	1.18c
Earlyana	3.69a	3.60a	3.48a	1.33a	1.19b	1.09b
MG II						
Amsoy	3.27a	2.40b	1.84c	1.25a	0.94b	0.73c
Lindarin	2.84a	1.94b	1.78b	1.02a	0.69b	0.60b
Magna	4.40a	3.66b	3.20c	1.93a	1.54b	1.30c
Mukden	3.44a	2.94b	2.80b	1.20a	1.02b	0.92b
Seneca	3.43a	2.92b	2.80b	1.15a	0.93b	0.83b
MG III–IV						
Adelphia	3.16a	2.38b	1.98c	1.11a	0.74b	0.65b
Chief	2.82a	1.99b	2.03b	0.96a	0.66b	0.68b
Clark 63	3.62a	3.14b	2.50c	1.50a	1.23b	1.01c
Dunfield	3.50a	2.87b	2.90b	1.22a	0.95b	1.02b
Manchu	3.38a	2.72b	2.52b	1.02a	0.80b	0.74b
Overall mean	3.40a++	2.76b	2.48c	1.25a++	1.01b	0.89c

* Means (within cultivars) followed by the same letter do not differ at 0.05, LSD = 0.25 dm^2.

+ Means (within cultivars) followed by the same letter do not differ at 0.05, LSD = 0.12 g.

++ Overall mean values (within variables) followed by the same letter do not differ at 0.05, LSD = 0.06 dm^2 (LSA), 0.02 g (LDW).

Table 56

Mean photosynthesis per unit leaf area for individual cultivars measured at three soil water potential levels (Sammons et al. 1978)

Cultivars	Soil water potential in bars		
	−0.7	−3.0	−8.9
	($mg\ CO_2\ dm^{-2}\ h^{-1}$)		
MG 0–00			
Cayuga	14.8a*	9.2b	5.7b
Grant	7.9a	6.1a	8.3a
Manitoba Brown	10.0a	5.3b	0.7c
Pagoda	12.5a	5.8b	6.5b
MG I			
A 100	14.9a	10.0b	8.3b
Black Eyebrown	20.8a	20.2a	14.1b
Blackhawk	18.6a	8.6b	4.9c
Chippewa 64	15.3a	8.0b	5.3b
Disoy	8.6a	7.4ab	4.7b
Earlyana	9.2a	6.1a	5.6a
MG II			
Amsoy	20.2a	16.8a	12.4b
Lindarin	14.5a	13.1a	7.3b
Magna	15.0a	14.6a	11.3a
Mukden	16.2a	6.1b	5.6b
Seneca	9.4a	4.3b	2.1b
MG III–IV			
Adelphia	12.8a	8.7b	5.3b
Chief	11.9a	10.2ab	7.4b
Clark 63	14.1a	8.8b	8.6b
Dunfield	10.7a	7.8ab	6.6b
Manchu	14.8a	12.2a	12.4a
Overall mean	13.6⁺	9.5b	7.2c

* Means (within cultivars) followed by the same letter do not differ at 0.05, LSD = 3.7 mg $CO_2\ dm^{-2}\ h^{-1}$.

⁺ Overall mean values followed by the same letter do not differ at 0.05, LSD = 0.8 mg $CO_2\ dm^{-2}\ h^{-1}$.

There are varieties in which the intensity of photosynthesis fell to one-quarter at 8.9 bar, while in others—e.g. in Manchu—it hardly changed.

Many researchers have recently tried to produce drought tolerant types by making selection on the basis of the root development of varieties and strains. Bedrov (1975) found that the root development of varieties

Table 57

Varietal differences in lateral root development as a function of cultural practices (Bedrov 1975)

Variety	Cultural practices	Number of developing roots					
		1966	1967	1969	1970	1971	1972
Saratovskaya 29	irrigated	6.31	2.70	2.51	5.67	3.94	5.60
	dry	4.37	0.81	1.39	3.65	0.60	1.40
Saratovskaya 38	irrigated	6.37	2.51	2.90	6.05	3.34	5.60
	dry	4.70	1.36	2.04	3.47	2.97	1.60
Scala	irrigated	6.66	4.10	2.54	4.76	1.68	4.50
	dry	5.23	1.20	1.93	2.53	1.38	1.20
Sortandinka	irrigated	10.05	7.00	3.00	5.95	3.42	6.50
	dry	5.80	1.61	1.06	0,70	2.46	1.40
Chainik	irrigated	7.17	1.89	1.81	4.20	3.16	4.90
	dry	4.70	0.40	1.67	3.15	2.14	1.10
Merit	irrigated	6.76	3.92	4.87	5.59	3.27	8.00
	dry	5.29	1.08	1.60	3.30	1.71	1.70
Tera	irrigated	9.97	4.81	2.50	4.60	2.49	5.90
	dry	6.50	2.37	2.38	2.62	1.65	2.20
Norrena	irrigated	9.46	4.57	—	3.46	2.58	4.90
	dry	6.02	0.95	—	3.42	0.27	2.20

produced in drier regions (east of the river Volga, Kazachstan) was more intensive, and the radicle and the whole root system developed more vigorously (Table 57) than e.g. in the Skandinavian or Canadian varieties.

Drought tolerant varieties showed no essential differences in laboratory experiments under normal and dry conditions. Among plants selected from hybrid populations types developing 5 branch roots exceeded by 11.7–50 per cent those with 3 branch roots. Yield surplus was produced by the progeny in 3 years, and the ability to develop 5-branch roots became stabilized. A type with 6-branch roots appeared in 1–3 per cent.

The data presented suggest—for the time being without the conditions of inheritance known—that it is possible to produce cultivars with better water utilization and higher resistance to drought.

CHAPTER 5

Genetic effects of herbicides and pesticides

The wide application of herbicides and pesticides called attention to the necessity of studying their possible side-effects with a view to preventing losses and protecting the environment (Nagy 1979).

Effect of pesticides on the composition of natural populations

Herbicides have a remarkable effect on the composition of natural plant stands, manifest first of all by changes in the structure of weed populations. The changes are due partly to the genuine resistance of certain species to the chemicals applied, partly to the fact that owing to the one-sided use of herbicides from species originally sensitive to them resistant types are selected and spread (Újvárosi 1969).

Of the species wide-spread in Hungary in the fifties such weed species as *Bromus secalinus, Papaver rhoeas, Sinapis arvensis, Polygonum aviculare* etc. have been controlled by the leaf and root herbicides, while resistant species e.g. *Galium aperine, Convolvulus arvensis, Galeopsis tetrahit* etc. have become numerous. Owing to the use of amino-chloro-triazine type of herbicides *Avena fatua, Apera spica-venti* have spread in cereals; *Agropyron repens, Cynodon dactylon, Echinochloa crus-galli* in hoed plants, *Sorghum halepense* (in maize) etc.

According to Ubrizsy (1972) under the influence of these chemicals (remaining active for 4–5 years in the soil, and the maize and vine monocultures increase their effectiveness) a negative selective transformation of the weed flora began in the sixties and species forming the original weed population have been replaced by those tolerant or resistant to herbicides.

With respect to these weed species, the literature generally discusses the selection of resistant and tolerant types, and whether these herbicides have a mutagenic effect, but whether the process—beside the selection of resistant types—should not be regarded as an *organic selection* is not even raised. In the course of organic selection, the organism probably adapts itself genetically to special conditions (in the present case to herbicides) by undergoing first a non-heritable modification resistance. This then becomes stabilized in a process of selection through favourable mutants arising in the meantime.

Mutants may be induced in the living organisms of the soil too. Singh et al. (1979) in India studied the mutagenic effect of alachlor (Lasso) used for the weed control of rice fields on the nitrogen-fixing blue green alga species *Nostoc muscorum*. It was compared with the supermutagen N-methyl-N'-nitro-N-nitrosoguanidine and proved to be more effective on Nostoc. To test the mutants the appearance of MA-R (methyl-amine resistance) was used. The mutants—unlike the initial material—are able to make use of methyl-amine as a nitrogen source, but at the same time inhibit photophosphorylation. The rate of spontaneous appearance of MA-R mutants is $2.5 \pm 2.4 \times 10^{-6}$, that of alachlor induced ones $7.6 \pm 6.8 \times 10^{-2}$. The herbicide concentration applied in the experiment was not higher than the one affecting the blue-green alga in the soil. In the cultures 55 per cent of the algae remained alive.

As a final result, forms defective in growth and reproduction were produced. Through N-fixation the blue-green algae greatly contribute to the maintenance of soil fertility, to the nitrogen supply of the soil, therefore the multiplication of defective mutants is economically disadvantageous.

The application of herbicides may change the composition of stands even when the species in question are not equally sensitive to the herbicides applied. Triplett et al. (1977) studied the effects of Pronamide (3,5 dichloro (N-1,1-dimethyl-2-propynyl)-benzamide), Simazine (2-chloro-4,6-bis (ethylamino)-s-triazine) and 2,4 D (2,4-dichlorophenoxy) acetic acid) in an artificial grassland consisting of lucerne, orchardgrass, Kentucky bluegrass and dandelion. The use of 2,4 D did not change the ratio of lucerne but reduced the share of dandelion. Pronamide (0.55 kg/ha) and Simazine (1.1 kg/ha) modified the composition of the stand in the way seen in Fig. 34.

According to the figure the share of lucerne within the stand grew at the expense of the grass species, of Kentucky bluegrass in the first place. The experiment suggests that the change in composition of a stand depends on the herbicide sensitivity of species (and varieties) composing the stand.

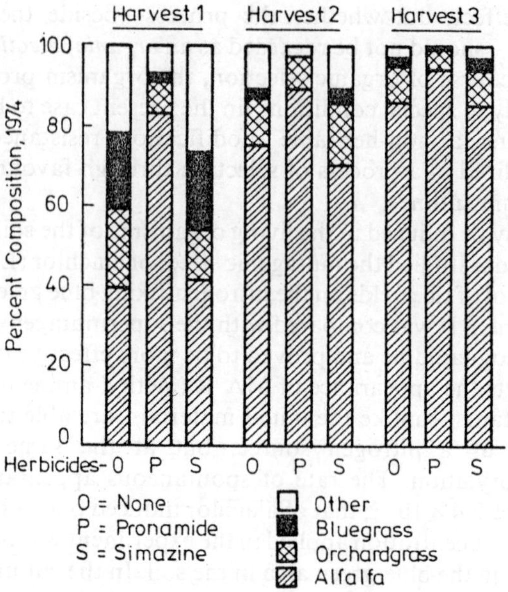

Fig. 34. Forage composition of individual harvests in 1974 as influenced by the herbicides Pronamide and Simazine (Triplett et al. 1977)

Mention should be made of a less frequently studied aspect of pesticide application: of the effect of seed dressing too. After seed treatments with 3 fungicides, 4 insecticides and their combinations Grogan et al. (1959) followed the action of these chemicals during storage (for 3, 6, 12 and 18 months, respectively) on the double-cross maize hybrid Kansas 1639. The summarized results of experiments are shown in Fig. 35.

The most balanced effect was obtained with the fungicide–insecticide combination, while the insecticides when applied by themselves had the least favourable effect. Besides the decrease of germinability it would have been worthwhile to determine yield averages as well. The experiment ought to be repeated with recently used pesticides and a larger number of hybrids.

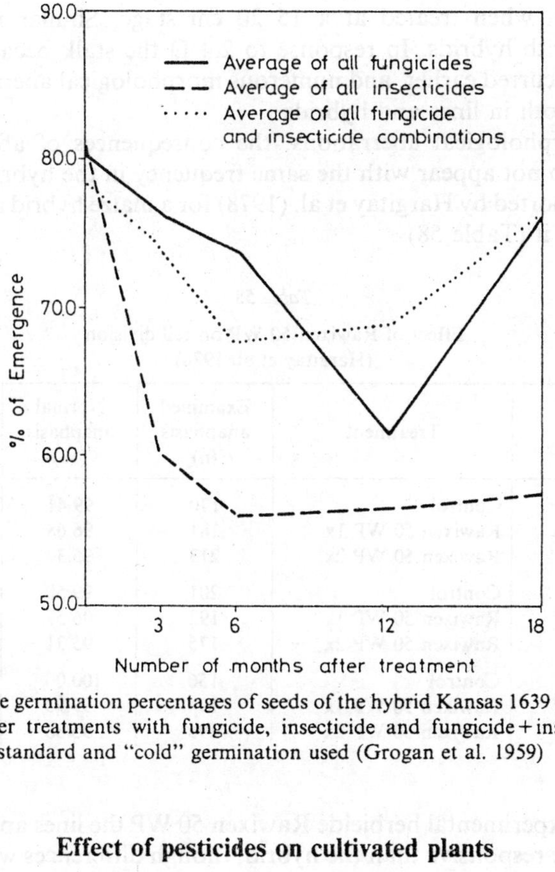

Fig. 35. Average germination percentages of seeds of the hybrid Kansas 1639 0, 3, 6, 12 and 18 months after treatments with fungicide, insecticide and fungicide + insecticide with standard and "cold" germination used (Grogan et al. 1959)

Effect of pesticides on cultivated plants

To know the behaviour (responses to chemicals) of varieties, lines, hybrids is of practical importance too, in consideration of modifications and possible mutations.

Research work of this type began already in 1949 with maize lines (Jugenheimer 1958); 145 lines were tested for sensitivity to 2,4 D. The most resistant lines were marked *2*, the most sensitive ones *5*. Mark *2* was given to 12 lines, *3* to 34, *4* to 70 and *5* to 26 lines.

Miller (1958) studied the effects of 0–0.5–1 pound of 2,4 D on 19 maize lines and 8 hybrids under field conditions in Minnesota. Increasing rates of 2,4 D caused a proportionate reduction of yield. Some lines responded with a much sharper yield reduction to 2,4 D applied at the 45–50 cm height than when treated at a 15–20 cm stage. Similar results were obtained with hybrids. In response to 2,4 D the stalk became frail the ripening occurred earlier, and numerous morphological aberrations were observed both in lines and hybrids.

The morphological aberrations, the consequences of abnormal cell division, do not appear with the same frequency in the hybrids or in the lines, as reported by Hargitay et al. (1978) for a maize hybrid and the lines composing it (Table 58).

Table 58

Effect of Rawixen 50 WP on cell division
(Hargitay et al. 1978)

Line and variety	Treatment	Examined anaphasis (n)	Normal anaphasis (%)	Abnormal anaphasis (%)
	Control	170	99.41	0.59
N6	Rawixen 50 WP 1x	181	96.68	3.32
	Rawixen 50 WP 2x	213	95.30	4.70
	Control	201	99.50	0.50
WF9	Rawixen 50 WP 1x	192	96.57	3.43
	Rawixen 50 WP 2x	175	95.31	4.69
	Control	156	100.00	0.00
MVSC-620	Rawixen 50 WP 1x	194	97.93	2.07
	Rawixen 50 WP 2x	195	95.38	4.62

To the experimental herbicide Rawixen 50 WP the lines appeared to be much more responsive than the hybrid, though differences were found in herbicide sensitivity between the two lines too.

Kádár and László (1975) have reported on the herbicide sensitivity of domestic and foreign hybrids grown in Hungary (Table 59).

This table enables the herbicide sensitivity to be assessed under field and glasshouse conditions, on various types of soil and with different rates of herbicide application. There are hybrids, e.g. H-250, DeKalb-238, SLX-31g, that exhibit high responsiveness in most cases, while others (MVSC 580, SLX-290) are sensitive only to larger doses. It would have been interesting to study the effect of herbicide treatments on the volume of yield as well.

Pap et al. (1977) examined the effect of various concentrations of Igrán 80 WP (4-ethyl-amino-2-*t*-butyl-amino-6-methylthio-s-triazine) on

Table 59

Result of testing maize varieties for herbicide sensitivity
(Kádár and László 1975, abridged)

Maize variety	Field trial, dose (kg/ha)								
	5.0	10.0	20.0	1.5	3.0	6.0	1.5	3.0	6.0
	Buvinol			Maloran			Afalon		
Anjou-255	0	0	+	0	0	0	0	0	+ +
Anjou-360	0	0	+	0	0	0	0	0	+ +
DeKalb-238	+ +	+ +	+ + +	+	+	+ +	+	+ +	+ + +
DKXL-22-B	0	0	0	0	0	0	0	0	+
DKXL-316	0	0	+	0	0	0	0	0	+ +
DKXL-342	0	0	0	0	0	0	0	0	+
Funk G-4465	0	0	0	0	0	0	0	0	+ + +
Funk G-4366	0	0	+	0	0	+	0	0	+ + +
Funk G-4384	0	0	0	0	0	+	0	0	+ + +
Funk G-4195	0	0	0	0	0	+	0	0	+ + +
H-5	0	0	+	0	0	+	0	0	+ + +
H-6	0	0	+	0	0	+	0	+	+ +
H-11	+	+	+	+	+ +	+ +	+	+ +	+ + +
H-15	0	0	0	0	0	+ +	0	0	+ +
H-250	+ +	+ +	+ + +	+	+ + +	+ + +	+ +	+ + +	+ + +
H-350	0	0	+	0	+	+	0	+	+ + +
H-457	0	+	+	0	+	+	+	+	+ +
MVSC-380	0	0	0	0	0	0	0	0	+ +
MVSC-405	0	0	0	0	0	0	0	0	+ +
MVSC-429	0	0	0	0	0	+	0	0	+ +
MVSC-580	0	0	+ +	0	+ +	+ + +	0	+ +	+ + +
MVSC-598	0	0	0	0	0	0	0	0	+ +
OSSK-212	0	0	0	0	0	0	0	0	+ +
OSSK-218	0	0	0	0	0	0	0	0	+ +
OSSK-440	0	0	0	0	0	0	0	0	+ +
OSSK-485	0	0	0	0	0	+ +	0	0	+
PX-20	0	0	0	0	0	0	0	0	+
PX-50	0	0	0	0	0	0	0	+	+ +
PX-442	0	0	0	0	0	0	0	0	+
PX-448	0	0	0	0	0	0	0	0	+
SLX-20	0	0	0	0	0	+ +	0	0	+ + +
SLX-29-c	0	+	+ +	0	+	+ + +	+	+	+ + +
SLX-319	+	+ +	+ + +	+ + +	+ + +	+ + +	+ + +	+ + +	+ + +

Soil type: Sandy soil.
Damage: 0 = symptomless; + = slight; + + = medium; + + + = serious.

nitrate reductase activity in 4 hybrids in the field, in the lines of a single-cross hybrid, and in 10-day-old seedlings of the F_1 and F_2 generations.

In the experiment performed with four hybrids a 1–2-day delay of flowering in 2 hybrids, a general reduction of plant height and varying extents of dry matter accumulation in the leaves were observed (Fig. 36).

A comparison of these data to those on the *in vivo* NR-activity of maize plants grown in the field (Table 60) reveals the stimulative effect of pre-emergent Igrán treatments on nitrate reduction in the leaf tissue of maize. A dose of 3 kg/ha increased the value of NR-activity in three hybrids by 4.7, 2.8 and 3.43 per cent, respectively, compared to the control. In hybrid No. 4 no N stimulation could be pointed out with the same rate of Igrán treatment.

A quantity of 6 kg/ha considerably raised the value of NR-activity (1) in all hybrids, particularly in hybrid No. 3 (52.3 per cent). The lowest effect of pre-emergent Igrán applications on nitrate reduction was attained in hybrid No. 4 (19.5 per cent).

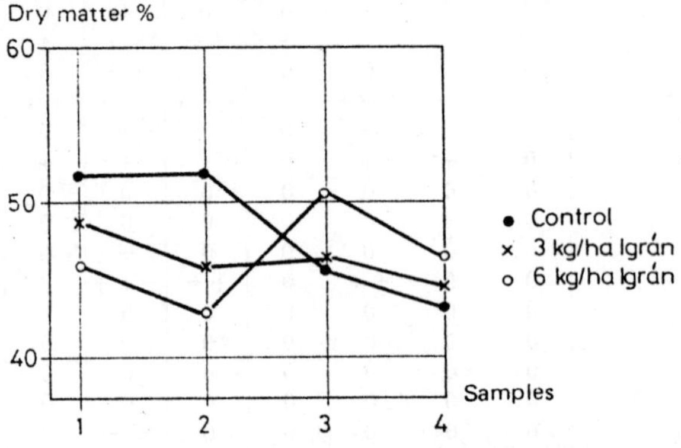

Fig. 36. Effect of pre-emergent application of Igrán on the dry matter accumulation of maize leaves in the field (Pap et al. 1977)

The values of NR-activity related to dry matter (2) differ from the above results. In hybrids No. 1, 2 and 3 the nitrate reduction was stimulated (by both doses) in this case too; in hybrid No. 4, on the other hand, the values of NR-activity (2) decreased irrespective of the rate of application.

Dry matter contents determined simultaneously with the nitrate reduction measuring indicate that in hybrids No. 1 and 2 the stimulation of NR did not involve an increase in the dry matter content of the leaf tissue, although the NR-activity (2) increased by 57.8 and 67.7 per cent, respectively (in response to a quantity of 6 kg/ha applied). In these hybrids the treatments resulted in a decrease in the dry matter content of the leaf tissue. Hybrid No. 3 showed an increase in the dry matter content beside the values of NR-activity (2) increasing in accordance with the rate of

Table 60

Effect of pre-emergent Igrán applications on the *in vivo* NR activity of maize plants grown in the field (Pap et al. 1977)

Treatments	Hybrids	NR (1)*	NR (1)** activity, %	NR (2)*	NR (2)** %
Control	No. 1	0.97	100.0	1.85	100.0
	No. 2	1.02	100.0	1.96	100.0
	No. 3	0.97	100.0	2.14	100.0
	No. 4	1.00	100.0	2.78	100.0
3 kg/ha Igrán	No. 1	1.02	104.7	2.10	113.5
	No. 2	1.05+	102.8	2.30	117.3
	No. 3	1.31	134.3	2.84	132.7
	No. 4	1.00+	100.0	2.21	79.4
6 kg/ha Igrán	No. 1	1.34	137.9	2.92	157.8
	No. 2	1.31	131.7	3.17	161.7
	No. 3	1.48	152.3	2.85	133.1
	No. 4	1.19	119.5	2.59	93.1

* NR activity (1) = μmol nitrite/g fresh weight (hour produced). NR activity (2) = μmol nitrite/g dry matter/hour produced.

** The values show the changes of NR activity (1) and NR activity (2) as a percentage of the control.

+ = No significant difference in comparison to the control ($SD_{5\%}$).

herbicide application. The dry matter content of hybrid No. 4 increased to some extent (by 3.7 per cent with a dose of 3 kg/ha and by 6.9 per cent when the quantity of herbicide applied was 6 kg/ha) irrespective of the decreasing values of NR-activity (2).

Thus, the correlation is again non-linear, and there are differences depending on the variety too.

In order to become familiar with the action mechanism Hanikov et al. 1971 (in: Bálint 1977) studied the effect of 2,4 D in the sensitive *Sinapis alba, Fragaria grandiflora,* and other species, in *Phaseolus vulgaris, Vicia faba, Pisum sativum* and other less sensitive species as well as in perfectly resistant cultures. In the first group the herbicide moved from the leaves to the growth tip and stalk, in the resistant plants it remained in the leaves, while in the less sensitive group the flow of herbicide from the leaves was slower.

Subsequently the influence of various herbicides on the volume of yield on the one hand, and on the baking quality, on the other, in two wheat varieties can be observed. The examination results obtained by Pollhamer (1973) are contained in Figs 37 and 38.

According to the evidence of the data some of the herbicides applied considerably decreased the yield. The trend of the quality index is independent of this indicating the physiological effect of the herbicides in this case.

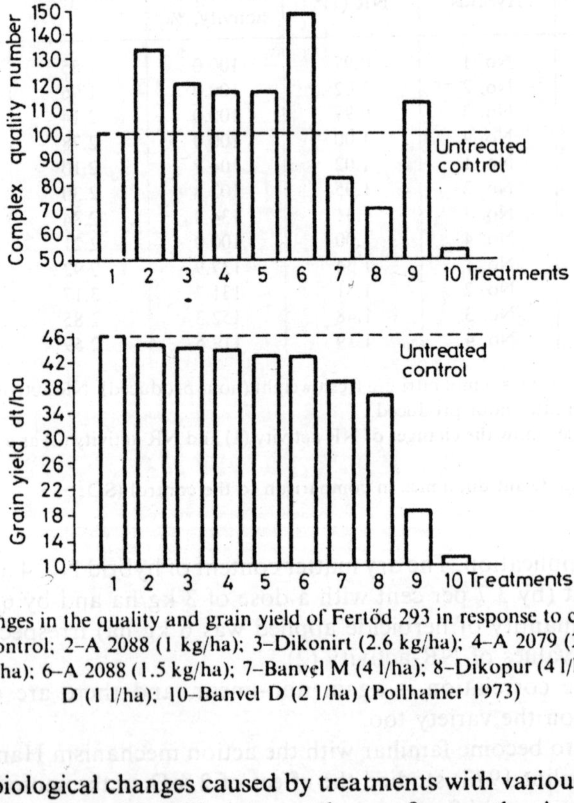

Fig. 37. Changes in the quality and grain yield of Fertöd 293 in response to chemical weed control: 1–control; 2–A 2088 (1 kg/ha); 3–Dikonirt (2.6 kg/ha); 4–A 2079 (2 kg/ha); 5–A 2079 (1.5 kg/ha); 6–A 2088 (1.5 kg/ha); 7–Banvel M (4 l/ha); 8–Dikopur (4 l/ha); 9–Banvel D (1 l/ha); 10–Banvel D (2 l/ha) (Pollhamer 1973)

Of the biological changes caused by treatments with various chemicals the reduction of germinability occurred most frequently. Among the 15 pesticides tested Banvel D and Lorox had a particularly strong inhibitory effect on germination. Lorox caused, in addition, chlorophyll deficiencies and a one month delay of shooting up. In experiments with barley (Stroev in: Badawi 1973) no similar effect was observed (possibly due to the lower concentration used for the seed primordia). One of the pesticides (IPA) caused a significant reduction of height; the same pesticide decreased the ability for tillering, while another (Simazine) increased it. MAH increased

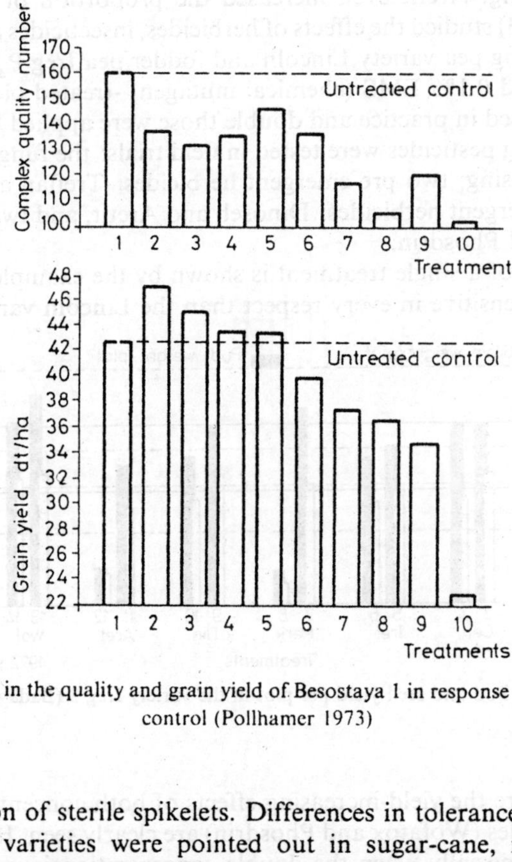

Fig. 38. Changes in the quality and grain yield of Besostaya 1 in response to chemical weed control (Pollhamer 1973)

the proportion of sterile spikelets. Differences in tolerance to pesticides between the varieties were pointed out in sugar-cane, rye-grass and vegetable crops. Owing to differences in tolerance the herbicides have a varying effect on photosynthesis and respiration—as suggested by the results of investigations made by Romanovskaya et al. (1968) with phenazone. In sensitive species (bean, cabbage, spurrey) the CO_2 assimilation decreases at a fast rate, while in resistant species—like the sugar-cane—a temporary decrease is followed by the restoration of the normal level, or even some stimulation is observed.

Beside a direct yield-decreasing effect shown in various plant species (lettuce, flax, potato etc.) Coiteux and Jassmin (1964, in: Badawi 1973) found the volume of yield to decrease after five years of treatments with the herbicides Garlon, Monuron and Amitrol compared to the hand-hoed control. In pea the treatments increased the number of flowers and extent of seed setting. Aretit even increased the proportion of root-nodules. Badawi (1973) studied the effects of herbicides, insecticides and fungicides on the canning pea variety Lincoln and fodder pea Ireg P_3. As a control untreated and 0.1% EMS (chemical mutagen) -treated plots were used. Quantities used in practice and double those were applied by the author. The following pesticides were tested in field trials: the fungicide Ceredon for seed dressing, two pre-emergent herbicides: Treflan and Merkazin, two post-emergent herbicides: Dinoseb and Aretit, and two insecticides: Wofatox and Phosdrin.

The effect of a single treatment is shown by the example of Ireg P_3, a variety less sensitive in every respect than the Lincoln variety (Fig. 39).

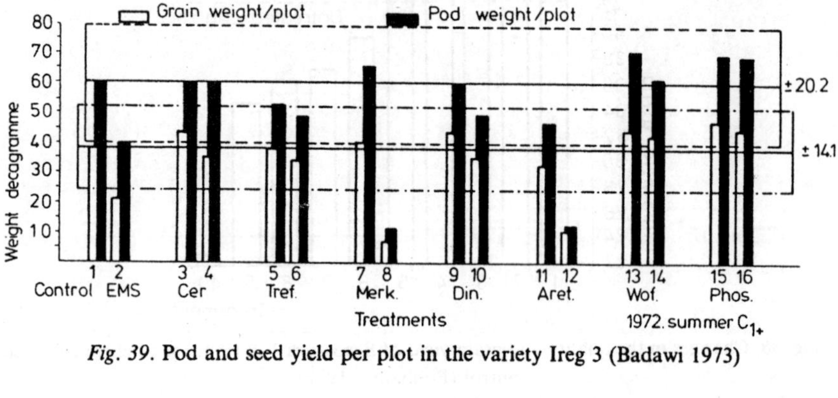

Fig. 39. **Pod and seed yield per plot in the variety Ireg 3 (Badawi 1973)**

In the figure the yield-increasing effects of both concentrations of the two insecticides (Wofatox and Phosdrin) are clearly seen. However, yield reduction (especially when the double concentration is used) is a more usual consequence (in the case of 5 pesticides!). The effect of Aretit is remarkable. The pod weight per unit area shows a similar–though not always a linear–tendency.

Data of the literature and the experiments conducted by Badawi (1973) at the Gödöllő University of Agricultural Science suggest that not only overdoses but even the normal doses of certain chemicals may cause unexpected yield reduction. A modification like that may be considerable when herbicide sensitive varieties or combinations are used.

The defective physiological changes (chlorophyll deficiency, teratologic morphological changes, sterility etc.) are consequences of irregularities in cell division which are characterized by chromosome number in aberrations but in other ways, too.

Besides the physiological damage defective mutative changes as a genetic consequence may also occur if the aberrant cells have a part in seed setting.

Nygren (1949) gave account of herbicides, including 2,4 D and 2,4 T, causing cytological deficiencies. Chromosome aberrations caused by herbicides were found by a number of authors to be similar to changes due to chemical mutagens, colchicin and radiation. Chromosome breaks, bridges, sticking, aneuploid and polyploid complements etc. have all been encountered. Grant (1964) observed aberration of 3–19 per cent after seed dressing with 6 herbicides, and only 9 per cent after X-ray treatments. *Vicia faba* seedlings treated with 15 pesticides had 9.2–21.3 per cent aberrant cells in their root tips. The inhibition and disturbance of meiosis and mitosis resulted in sterile pollen grains and other morphological defects. The cytological effect of insecticides (Sevin, Metapa, Dimecron 110 etc.) in several cases was similar to that of EMS. Of the fungicides those containing lead were found already in 1937 (Sax) and 1939 (Kostoff) to cause aberrations. With the pesticides used at present (Ferbam, Botran etc.) some authors have obtained similar results; Tomkins and Grant (1972), on the other hand, observed no aberration in the root tip cells of barley seedlings when using tetra-chloro-iso-flavonitryl (DAC).

Post-effect studies

High rates of fertilization may also increase the proportion of chromosome aberrations to a considerable extent. In an experiment (Hajós-Novák and Bálint 1978) we studied the proportion and type of chromosome aberrations in maize as a response to two different rates of fertilizer application, in the year of treatment and one year later (post-effect without treatment).

In the year of the treatment samples were taken from ears of the free-pollination hybrid maize K SC 360 for cytological examinations. Sample 1 was taken from the plot given 380 kg/ha NPK active agent, while sample 2 from the one treated with 1180 kg/ha active agent. The ratio of nutritive elements was 1 : 0.9 : 0.7 in both treatments. The amount of precipitation during the vegetation period was 400 mm.

In the year following the treatment the area did not receive fertilizers. We sowed what had been left from the previous year's sowing in microplots arranged in random block design with four replications on the same area, and took the samples from plants thus obtained.

The aberrations were studied in meta- and anaphasis in 1,000–1,800 cells per treatment. Aberrations in the metaphase were divided into two groups: translocation and other aberrations. The latter included divided metaphase, peripheral chromosomes, fragments, arrhythmic separation, and adhering chromosomes.

The results of cytological examinations performed in the year of the treatment are shown in Table 61. As seen from the data, the lower rate of nutrition resulted in 2.9 per cent, the higher in 2.8 per cent of aberration compared to 0.5 per cent in the control. The proportion of aberrations showed the same trend in the two treatments and was about six times as high as in the control.

Aberrations in both treatments occurred most frequently in the metaphase, and then in the form of other aberrations. As regards aberrations in the anaphase the situation was less favourable in the 1180 kg/ha NPK treatment. In the year following the treatment we repeated the analysis of mitosis—without any further treatment applied—in the same way as before. We wanted to know whether or not the number of aberrations would decrease. The number of aberrations was reduced significantly from 2.9 to 1.6 per cent with the lower dose, and unreliably from 2.8 to 2.3 per cent in the larger dose. This corresponds to a decrease of 1.3 and 0.5 per cent, respectively. Differences in the number of chromosomes compared to the normal chromosome complement were not observed.

Wuu and Grant (1966) studying the proportion of aberrations in the progeny found much less chromosome aberration in the C_1 or C_2 (than in the C_0) generation of barley grains treated with 9 herbicides; they never exceeded 2 per cent. Insecticides Endron and ENT-50612 gave similar results only a low percentage of aberrations observed in C_1 appeared in the following generations. This phenomenon can be explained by the lower rate of division of defective cells compared to the normal ones, whereby the healthy cells outnumber the defective forms, fewer of which thus participate in the reproduction process. In short: in the course of natural selection only the viable forms survive. The idea that the repair mechanism

Table 61

Number of chromosome aberrations in mitosis in response to different rates of fertilization in the year of treatment (Hajós-Novák and Bálint 1978)

Treatment	Metaphasis					Anaphasis				Total abnormal, n %
	Normal, n	Chromosome fragments, n	Chromosomes stuck together, n	Arithmic separation, n	Translocation, n	Normal, n	Bridge, n	Chromosome fragments, n	Other, n	
380 kg/ha mixed active ingredients	1,094	23	15	2	2	390	1	—	2	45 = 2.94
1180 kg/ha mixed active ingredients	1,531	25	15	1	—	401	5	3	7	56 = 2.81
Control	883	4	—	—	1	318	—	1	—	6 = 0.50

D ∅–1 = 2.44%
D ∅–2 = 2.31%
D_α = 1.58%
D = difference between the control and the aberration percentage of the treatment in question
D_α = calculated smallest reliable difference
SD = 0.01%.

is in operation may arise, but no experimental evidence is available at present to support this.

As opposed to the eliminative process described above there are data on the maintenance of aberrations over several generations. After treating tomato seeds with malein-hydrazide solution chromosome aberrations were found in the third generation. In barley and *Tradescantia,* Tomkins and Grant (1972) pointed out a higher rate of mutation in the progenies than in the control. As a result of treatments with 2,4 D, Dark (1967) obtained a stable tomato variant producing plum-shaped fruit in the third generation from the treated round-fruited type.

These data suggest that the cytological effects of treatments decrease in the progeny, and are maintained only in a few cases. No significant increase in the rate of mutation occurs, which is very important in the agricultural practice.

Badawi (1973) carried out cytological analyses of the effects of already known chemicals and studied the productivity in the progeny. The rate of mutation did not increase, but differences appeared in the M_2 generation in chromosome aberration, productivity and pod formation alike.

Morphological mutants

To point out the mutagenic effect in maize a simple technique was used. In the progeny of sweet corn (su_1 su_1) produced with the pollen of a mutagen treated normal (Su_1 Su_1) parent, the mutations can be tested well through the double recessive grains. Table 62 shows the effect of the herbicide Rawixen 50 WP (Hargitay et al. 1978).

On the basis of the data it can be established that the mutation rate induced in the pollen by Rawixen 50 WP ($Su_1 \rightarrow su_1$) grew parallel with an increase in the dosage rate both in the lines and in the hybrids formed by them. The highest sensitivity was shown by the line N6. The higher sensitivity of the parent line N6 appeared in the MVSC-620 hybrid too. Sensitivity was lower in line WF 9 than either in the MVSC-620 hybrid or in the N6 line.

Hargitay (1976) examined the inbred maize line WF9 for its response to Hungazin PK treatments. The experiment consisted of the following treatments: a single dose of 2.8 kg/ha, a 6-fold dose of 16.8 kg/ha and a control hoed twice. The treatments did not cause changes in the plant height and yield, but in the third generation following the treatment (M_3) morphological changes shown in Table 63 appeared.

Table 62

Effect of Rawixen 50 WP on the composition of maize kernels
(Hargitay et al. 1978)

	Treatment	Number of ears	Number of kernels examined	Number of sweet grains	Frequency of mutation, %
	Control	26	3,854	0	0.00
N6	Rawixen 50 WP 1 ×	22	3,277	5	0.1525
	Rawixen 50 WP 2 ×	24	3,164	7	0.2213
	Control	23	3,386	0	0.00
WF9	Rawixen 50 WP 1 ×	27	3,960	2	0.0505
	Rawixen 50 WP 2 ×	25	3,641	3	0.0824
	Control	26	3,621	0	0.00
MVSC-620	Rawixen 50 WP 1 ×	23	3,127	3	0.0959
	Rawixen 50 WP 2 ×	24	3,779	5	0.1323

Table 63

Phenotypic changes in the line WF9
(Hargitay 1976)

Treatments	Total number of plants	Dwarfness (n)	Leaf shape deformation (n)	Tassel deformation (n)	Total	
					(n)	(%)
WF9	318	1	—	—	1	0.31
HPK 1 ×	320	1	1	—	2	0.63
HPK 6 ×	316	4	2	1	7	2.19

As seen from the data of the table it was the dwarf form that occurred with the highest frequency. A comparison of the results of treatments reveals that the six-fold dose of Hungazin PK had the greatest effect. Morphological changes observed in the M_3 ran parallel with the increase in the rate of herbicide doses, that is, the effect of the chemical became stronger in the third generation and manifested itself in phenological changes.

The increase in the rate of mutation observed in the present case and generally expected in practice is very slight (recessive mutants are mostly induced), on the other hand, at certain stages of seed production it is possible to select the unfavourable variants. Attention must, however, be called to the danger of causing a large number of defective mutants to occur by using higher concentrations of these chemicals for reasons of technical error or carelessness, and thereby reducing the yield.

Thus the prescribed dosages must be observed not only for the economic implications but for environmental protection considerations as well. For changes in environmental conditions, the varied pesticide sensitivity of cultivars and lines, the different physiological and unknown genetic effects of chemicals call for continued studies on the variety–pesticide interactions, to avoid possible damage in practice. These studies must include considerations of potential genetic consequences.

CHAPTER 6

Structure of the ideal plant type

Demands on the ideal varieties have been formulated by an increasing number of authors (Baldy 1974, Austin and Jones 1975, Kurnik 1976). In the following sections the breeding tasks will be surveyed from the physiological genetics aspect.

We start again from the model of Wallace et al. (1973) (Fig. 4) and obviously have to try to increase the intensity of constructive processes, reduce the losses, and have as large a proportion of the produced organic matter incorporated in the crop as possible.

Some physiological parameters

Photosynthesis

In Chapter 2 several examples were given of the differences existing between the species in the intensity of photosynthesis. This process is the basis of organic matter production, still it does not by itself decide whether the yield will be large or small. Stoy (1976) used barley in his experiments to point out that the intensity of photosynthesis was essentially higher in the primitive than in the improved varieties. Planchon (1974) did not find as wide differences as that in his experiments with wheat (Table 64); the values are rather close to one another.

In comparison to *T. monococcum* the leaf area is in each case larger, but the tetra- and hexaploid wheats no longer show this difference.

The intensity of photosynthesis is connected with the gas exchange of the leaf. Stomatal resistance decreases the gas exchange thereby reducing the yield. According to the data of Peet et al. (1977) the stomatal resistance is inversely proportional to the volume of yield (Tables 65 and 66).

The above relationship is not uniform either in the successive phases of development or in the different varieties. Here mention should be made of the photoperiodic responses of varieties.

Table 64

Potential net assimilation in various species of the *genus Triticum* (Planchon 1974)

Species and botanical variation	Genome	Net assimilation⁺ (mg CO_2 $h^{-1}dm^{-2}$)	Leaf area (dm^2)
T. monococcum L.	AA	21.5	
-*var. Eredvianum* ZHUK.		—21.0	0.184
-*var. Hornemanni* CLEM.		—22.0	0.200
T. dicoccum SCHÜBL.	AA BB	20.7	
-*var. farrum* KRCK.		—21.0	0.443
-*var. pycnurum* KRCK.		—20.5	0.346
T. durum DESF.	AA BB	21.5	
-*var. hordeiforme* KÖRN. ("Lesnovo")		—21.5	0.300
-*var. leucurum* KÖRN. ("Montferrier")		—21.5	0.451
T. spelta L.	AA BB DD	19.2	
-*var. Arduini* KRCH.		—19.9	0.356
-*var. coeruleum* KRCH.		—18.5	0.495
T. aestivum L.	AA BB DD	18.8	
-*var. milturum* ALEF. ("Bladette de Puylaurens")		—19.5	0.360
-*var. lutescens* ALEF. ("Capitole")		—18.1	0.410

⁺ LSD between species 1.8
between varieties 0.7

Table 65

Growth and yield data for the nine varieties grown in 1973 and 1974 (Peet et al. 1977)

Variety	LAI⁺		Biological yield		Seed yield		Harvest index	
	1973	1974	1973	1974	1973	1974	1973	1974
			(kg/ha × 10^{-3})				%	
Black Turtle Soup	3.60	4.47	4.6	4.8	2.4	2.5	53	52
Michelite-62	2.11	7.33	4.2	5.2	2.0	2.3	46	44
Redkote	3.20	3.35	4.5	6.7	2.0	3.0	44	44
Red Mexican	2.11	1.59	3.6	5.8	2.0	3.8	56	66
Sanilac	1.94	4.15	4.1	7.1	1.9	3.9	47	55
Swedish Brown	1.37	2.82	4.3	7.4	2.4	4.7	55	62
Sutter Pink	2.38	4.67	3.0	3.5	2.1	2.3	70	65
Pinto	1.99	4.52	3.2	5.5	2.1	3.8	65	70
Redcloud	1.71	2.32	4.1	5.4	2.1	3.3	52	60
Average (mean across varieties)	2.27	3.91	3.95	5.7	2.1	3.3	52	58

⁺ Measured at first flowering.

Table 66

Photosynthesis (P) in mg CO_2 $dm^{-2}h^{-1}$ and stomatal resistance (r_s) in s cm^{-1} at different developmental stages in nine dry bean varieties and the percentage increase in P from flowering to early pod set (per cent increase) (Peet et al. 1977)

Variety	Mean across all developmental stage		First flowering		Early pod development		Per cent increase in P	Late pod development	
	P	r_s	P	r_s	P	r_s		P	r_s
Black Turtle Soup	17.48	2.09	11.81	1.64	14.27	2.31	21	15.31	2.10
Michelite-62	13.87	1.68	10.51	1.72	16.19	1.50	54	12.25	1.95
Redkote	13.66	2.38	7.61	3.62	17.66	1.67	132	17.81	1.48
Red Mexican	12.01	1.38	5.34	1.58	17.76	1.00	232	12.75	1.70
Sanilac	15.32	1.76	10.23	2.28	17.87	1.50	75	—	—
Swedish Brown	13.44	1.55	1.93	1.94	18.81	1.36	873	—	—
Sutter Pink	9.26	1.96	5.75	2.10	12.41	1.85	116	—	—
Pinto	4.99	1.78	2.72	2.10	6.11	1.62	124	—	—
Redcloud	10.20	1.59	2.81	1.68	14.31	1.54	408	—	—
Average (mean across varieties)	11.86	1.80	6.52	2.07	15.04	1.60	226	14.53	1.80
LSD (0.05)	3.07	0.32	1.77	0.70	5.14	0.41		5.00	0.47

The reason why Borlaug's (1972) wheat varieties could be introduced in 60 countries of the world is that they are indifferent to daylight.

Light demands must be taken into serious consideration when introducing foreign varieties because they may be the cause of failure.

Table 67

Effect of photoperiod.
Grain yield per plant (g) (Stoy 1976)

Genotype	Photoperiod		Quotient 15/24
	15 h	24 h	
CI 3947	4.14	1.58	2.62
CI 4362	4.47	2.56	1.75
CI 5195	3.81	2.12	1.80
Kristina	4.33	4.29	1.01

The effect of 15 and 24 hours of illumination on productivity and photosynthetic intensity is shown in 3 primitive barleys and a cultivated barley variety after Stoy's (1976) experiments (Table 67).

The varieties gave highly different responses to the two treatments. The local variety CI 3947 when exposed to light for 24 hours a day produced a smaller yield and showed a lower intensity of photosynthesis.

Chang and Vergera (1971) have provided a wider range of data on the effects of various day-lengths on the number of days required for shooting up in rice (Fig. 40).

In the tropics difficulties caused by the photoperiodic responses of varieties can be overcome by choosing the optimum sowing time; in Hungary possibilities for doing so are highly limited.

Respiration

As for the proportion of respiration losses early varieties have proved to be more favourable in some cases (Table 34).

Breeders have to take into account that respiration losses at night greatly depend on the temperature, so they can be reduced–particularly in the tropics–by selecting or breeding for higher tolerance to night temperatures. In fact, we only have indirect evidence as to whether or not it is really a case of reduced respiration losses.

The problem was studied in detail by Schaible (1962) in an experiment with 10 tomato varieties. The author examined the effect of night temperature on fruit setting and fruit size. The varieties Narcarlang and Porter yielded abundantly even at night temperatures of 23–28 °C. The

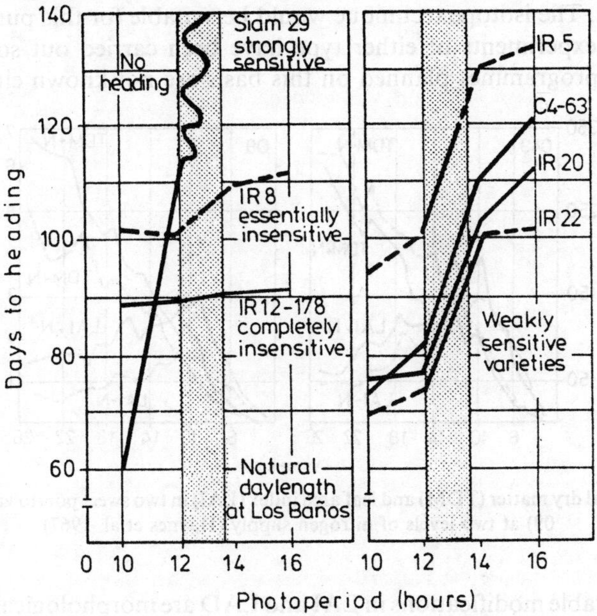

Fig. 40. **Effects of four photoperiod treatments from sowing to shooting in two rice varieties (Chang and Vergera 1971)**

commercial varieties Rutgers, Improved Garden State, Ace and Sioux while giving large yields at night temperatures of 14–20 °C produced very little when the temperature at night was 23–28 °C. The fruit size is reduced with higher night temperature in all varieties. Others by intercrossing varieties with low and high tolerance to night temperatures, respectively, obtained favourable combinations.

Of the complex factors of fruit setting the intensive loss of respiration is supposed to make its effect felt through the consumption of assimilates required for fruit setting.

Nutrient utilization

A better utilization of macro- and micronutrients would require the knowledge of the ratio of photosynthesis/respiration in the different varieties, and its trend at various levels of fertilization if genotypes more efficient than the existing ones are to be produced by crossing or mutagen treatment. The isotope technique would be suitable for this purpose, but very few experiments of either type have been carried out so far, and breeding programmes planned on this basis are not known either.

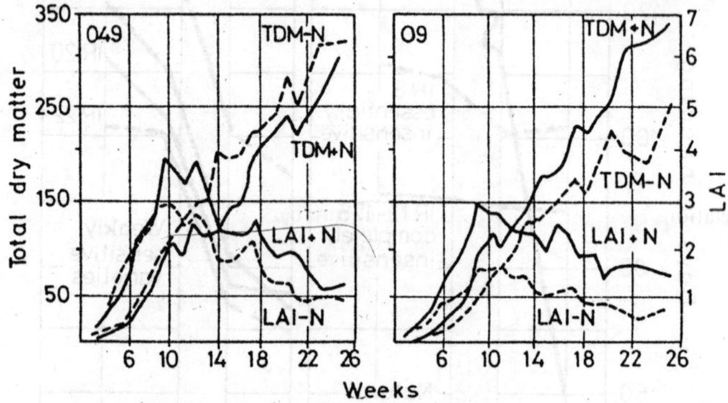

Fig. 41. **Total dry matter (TDM) and leaf area index (LAI) in two sweet potato varieties (049, 09) at two levels of nitrogen supply (Haynes et al. 1967)**

Favourable modifications in LAI and LAD are morphologically easy to detect. Such changes are, however, favourable only when resulting in an increase in the useful yield (HI). The responses of varieties are not uniform in this respect either, as seen from the investigation results of Haynes et al. (1967) (Fig. 41).

The experiments were performed with two sweet potato varieties (049 and 09). The nitrogen application increased the leaf area index in both varieties, but an intensive shoot growth in variety 049 delayed the development of the tuber, and after all the total amount of dry matter was therefore larger in the treatment not given nitrogen. The response given by the variety C9 met the expectations.

In plant species grown for their seeds the rate of fertilization and density of stand must be chosen so as to give the maximum possible yield. As a matter of fact, both factors may lower the value of HI, as shown by an experiment series (Fig. 42).

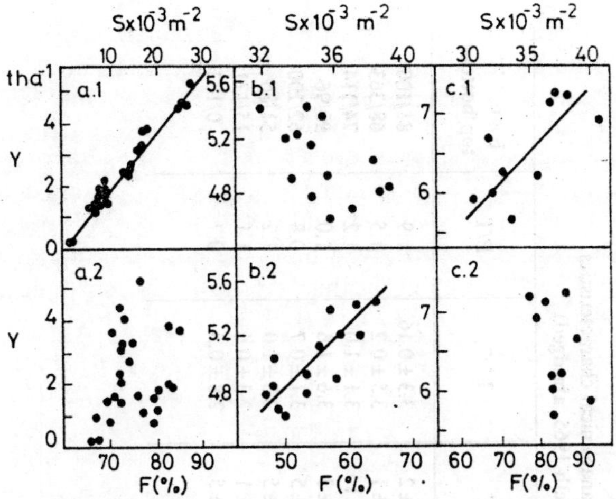

Fig. 42. Relations of grain yield, Y, to spikelet number, S, and the proportion of ripened grains, F, in rice plants cultured under: (a) various planting densities combined with nitrogen topdressing; (b) various combinations of ploughing depth, planting density, and nitrogen level; (c) different season with nearly optimum cultural conditions (Murata and Matsushima 1975)

The rice plants were cultured under: (a) various planting densities combined with nitrogen topdressing; (b) various combinations of ploughing depth, planting density, and nitrogen level; (c) different seasons with nearly optimum cultural conditions (Murata and Matsushima 1975).

In the case of a nitrogen topdressing (treatment (a)) the higher stand density resulted in a linear increase in the spike number and yield, but the proportion of mature grains was not as favourable as that. The effect of 3 factors (treatment (b)) was not reflected by a large yield in spite of a linear increase in fertility (proportion of mature grains); treatment (c) appeared to be the best from the point of view of both yield and fertility.

In countries of continental climate the varying conditions of precipitation also play a role.

Water supply

In evolving types utilizing the water well the lowest possible water consumption involved by an intensive assimilation may be the demand.

In this context data published by El-Sharkawy and Hesketh (1965) on the comparison of several plant species are of interest (Table 68).

Table 68

Photosynthesis, transpiration, CO_2 diffusion resistances, and anatomical characteristics of leaves of different species (El-Sharkawy and Hesketh 1965, abridged)

Species	Small veins, % area*	Large veins, % area*	Chloroplast diam., μ	No. of expts.	P+	T++	P/T	b × n,+++ top/bottom
Corn	6.5 ± 1.5	0	1.5	3	63 ± 2	3.3 ± 0.16	1.9	81/100
Sunflower	8 ± 1	0.5	3.0	4	50 ± 1	3.5 ± 0.2	1.5	68/363
Cotton (DPSL)	5.3 ± 1.5	0.6	2.5	2	38.1	3.1 ± 1.0	1.2	74/114
Oat	8 ± 1	0	1.5	3	31 ± 3	3.6 ± 1.6	1.0	90/96
Hibiscus tiliaceus L.	7 ± 3	1.6	—	3	23 ± 5	3.1 ± 0.7	0.8	62/200
Tobacco**	0 ± 0	1.6	—	5	21 ± 5	2.3 ± 1.0	1.2	51/86
Soybean	16 ± 3	0.6	3.5	2	25 ± 1	2.3 ± 0.6	1.1	15/121
Thespesia	13 ± 1	0.7	2.5	3	18 ± 5	2.8 ± 0.5	0.7	0/182
Cotton, PS–2	6.7 ± 1.5	0.9	3.5					
Bermudagrass	9 ± 1	0	1.5					

* No chloroplast cover.

\+ Photosynthesis in mg CO_2 $dm^{-2}h^{-1}$.

++ Transpiration in g H_2O $dm^{-2}h^{-1}$.

+++ b × n, the number of stomata x stomatal length per cm^2 leaf surface.

** *Nicotiana tabacum* L. var. Turkish Samson.

Table 69

Effect of temperature on the value of the transpiration coefficient (Petrasovits and Balogh 1975)

Plants	Air temperature	
	10–13°C	27°C
Plants of cold climates:		
Wheat	385	826
Barley	298	758
Oat	403	760
Lucerne	429	906
Plants of warm climates:		
Maize	249	210
Rice	2,566	585
Sorghum	1,236	223

The differences in the transpiration coefficient are essentially smaller than those of the intensity of photosynthesis. The best ratio of P/T (1.9) is almost three times as high as that of the lowest one (0.7). The data of Chapter 4 have convinced us that essential differences in the transpiration coefficient may exist even within the same species.

The transpiration coefficient is closely correlated with the temperature (Petrasovits and Balogh 1975, Table 69).

The data of the table suggest that in plants native to cold climates the transpiration coefficient is increased by high temperatures while in those of warm climates by low ones. It follows that a reduction or increase of heat demand–depending on the species–may result in improved water utilization.

This rule does not apply to all species. The transpiration coefficient of *Hevea brasiliensis* is lower at lower temperatures, but the latex production is at the same time higher (Shudherea 1971, in: Alvim and Kozlowski 1977).

Optimal models

After surveying some fundamental physiological paramaters, characteristic features of the optimal cultivars of various plant species will be described. Not only physiological but morphological characters are mentioned here, since–after all–each of them has something to do with the production potential.

Cereals

Physiological genetic model for wheat has been worked out by Austin and Jones (1975). The model though applied to the conditions of England contains many features which are of general validity (Table 70).

Moreover, the favourable and detrimental consequences are determined for each property.

According to Baldy (1974) the following properties should be contained in a wheat variety:

— Relatively small, upstanding leaves with the longest possible lifetime;

— Limited tillering (2–4 shoots), on flowering the spikes should form a "table";

— The last internode should be long (40–50 per cent of the total length of stalk);

— Long spike (more than 20 fertile spikelets), with 3 kernels per spikelet.

Kurnik (1976) having taken 40 experts' suggestions into consideration formulates the demands on the Hungarian wheat varieties as follows:

"60–80 cm long, not too thick stalk. Node of tillering deep set, lower leaves narrow. Length and width of leaves increasing upwards, leaves possibly upstanding until waxy ripeness. At a stand density of 500 plant/m^2, with a good capacity for nutrient and water utilization 1 kg dry matter should be produced with 250–300 litres water. Overall climatic and pathological resistance.

Number of kernel/spike: 50–60, thousand-kernel-weight: 50 g. Quality A_2, B_1, grain/straw ratio 1 : 1 or 1 : 0.8. Maximum fluctuation of yield 20–30 per cent.

And finally, the variety should fully meet the technological requirements (mechanization, milling and baking industry)".

Requirements formulated by Donald (1968) for the cereals in general are contained in Table 71.

Every model has common and individual features worth considering in the course of breeding.

The model constructed by Murata and Matsushima (1975) is different. The physiological factors for attaining maximum yield, the attainable volume of yield and the limiting factors for rice are analysed. The main points are given below.

The translocation of assimilates synthetised in rice to the grains begins two weeks before heading and lasts for four weeks afterwards. In high

Table 70
(Austin and Jones 1975)

Physiological attributes of the ideal model wheat plant for British conditions

Attribute	*Benefit*	*Disadvantage*
Large seed.	Rapid emergence of seedlings. Reduced sensitivity to depth of sowing.	Unknown.
High concentration of seed protein.	Rapid emergence and vigorous seedlings.	May be low yield of seed per hectare.
Adequate investment in root growth.	Seedlings do not become susceptible to drought.	Rapid depletion of soil water. Root growth may occur at expense of shoot growth.
Early and near-synchronous formation of tillers, absence of late tiller production. Following ear initiation, cessation of "surplus" tiller growth.	No competition from late-formed tillers which may have adverse effects on ear development in the early, fertile tillers. "Wasteful" use of water, minerals and assimilates is avoided.	Reduced ability to compensate for death of main shoot and tillers.
Homeostasis of tiller number per unit area of ground. In winter wheat, a suggested maximum number of ear-bearing tillers is 800/m².	Compensation for variation in seeding rates and seedling survival.	Plants may be able to "sense" competition from neighbours only after the formation of an excess of tillers so that homeostasis mechanism is dependent on tiller death.
Response to photoperiod and temperature to give very slow growth rates during cold spells and in the shortest days.	Favours development of winter-hardiness, provided that sensitivity of photosynthesis to low temperatures is less than that of growth, because intermediary metabolites, especially carbohydrates, accumulate.	Small plants, possibly with smaller ear primordia and fewer, or later developing, tillers.
Ability to take up and reduce nitrate rapidly; ability to store nitrate and reduced nitrogen compounds in excess of those required for immediate use.	Minimises loss of soil nitrogen by leaching and denitrification.	Accumulation of nitrate and reduced nitrogen compounds to toxic concentrations may occur at high levels of nitrogen fertilisation.
In spring wheats, response of ear development to photoperiod and temperature which will give homeostasis of spikelet numbers per	Yield per hectare can be maintained in late sowing by increasing seeding rate.	Need for information on how seeding rates should be adjusted to allow for variation in sowing date.

Table 70 (cont.)

ear over a wide range of sowing dates; coupled with this, a decrease in tiller production per plant with progressively later sowing.		
High rates of photosynthesis per unit leaf area. May be achieved by: (a) Low intracellular resistance to CO_2 uptake.	Increases potential for dry matter production. Increases efficiency of water use, especially important in drought.	May be compensating changes in leaf size and thickness which offset any benefits.
(b) Low photorespiration rate.	Particularly advantageous at high temperatures and in drought.	Unknown.
(c) Low stomatal resistance. High stomatal frequency.	Useful only in conditions of ample water supply.	In dry situations, there may be an unacceptably high rate of water loss.
Canopy structure to give the maximum interception of light and its optimal distribution over the leaves. This may entail selection of genotypes with erect upper and lax lower leaves.	Maximises canopy photosynthesis rates.	During grain filling, when the leaf area index has declined to 3 or less, light interception by green leaves may be less in the erect-leaved dwarfs, and canopy photosynthesis correspondingly reduced.
Highest possible dry weight per unit of cropped area at anthesis, the weight to be concentrated in the ear-bearing tillers.	Favours maximum uptake of nitrogen, and in turn, grain protein yield per hectare.	May be adverse competitive effects and ear development, and reduced efficiency of water use. May be difficult to obtain in genetic dwarfs. May be associated with late anthesis.
Stomata, sensitive to water stress, but reopening quickly after stress is relieved. Particularly relevant for spring wheats.	Prevents irreversible damage in periods of drought.	May be undesirable if periods of water stress are of only short duration.
High cuticular resistance to water loss.	Reduces transpiration rate. Particularly advantageous in drought.	Unknown.
Rapid export of nitrogen from leaves when their contribution to canopy photosynthesis is small.	Particularly important if soil nitrogen or water is limiting.	Unknown.

Table 70 (cont.)

Continued root growth during the critical phase of ear development and during grain growth.	Supply of water and nitrogen assured. (The presence of nitrate in the lowest parts of the soil profile explored by roots will favour uptake during this phase, especially in drought.) Continued uptake of soil nitrogen will delay or reduce the demand for nitrogenous compounds imposed on the upper leaves, and so prolong the period over which these leaves can supply assimilates for grain filling.	May be adverse competitive effects on ear development and grain growth.
Dwarf habit	Confers resistance to lodging.	May adversely affect crop microclimate and pattern of light interception. May be agronomic disadvantages.
High stem density (dry weight per unit stem length) and resistance to flexing.	Confers resistance to lodging. Stem reserves can, in part, be mobilised to provide "insurance" against possible shortfall in assimilate supply during the late stages of grain filling, caused by drought or disease.	May be adverse competitive effects on ear development. Too extensive mobilisation of stem reserves could lead to lodging and brackling.

yielding rice varieties the incorporation prior to heading is not considerable, the period of filling of kernels is longer. Let us assume a 40-day period of grain formation after heading and that a maximum amount of assimilates is available. Following from this maximum yield of rice can be planned as follows:

1 The average value of daily sunshine is 400 cal cm^2 of which 45 per cent is active.
2 Of radiation active from the point of view of photosynthesis 5.5 per cent is reflected from the leaves, another 10 per cent is absorbed by inactive tissues.
3 For the reduction of 1 molecule of CO_2, 8 photons are required, so the energy conversion is 26 per cent.

4 The saturation of the upper leaves by the 400 cal cm^2 sunshine causes a 17 per cent loss.
5 The conversion factor of dry matter production is 3900 cal g^{-1}.
6 The daily loss of respiration is estimated to be 1.5 per cent of the dry weight (a loss of some 1.5 kg/m^2), and it is 25 per cent of the gross photosynthesis.

Table 71

Some model characters for a cereal ideotype
(Donald 1968)

Morphological
Short, erect stem, resistant to lodging
Single culm
Short, erect, dark green leaves
Few small leaves
Physiological
Shade tolerance
Tolerance of high density planting at high rate of fertilization
High LAD* above flag leaf
Rapid translocation of sugars
High nitrate reductase activity
High interspecific competitive ability
Low intraspecific competitive ability
Reproductive
Early flowering and maturity
High harvest index
Large erect ears
Many florets per unit dry matter
High (nonlimiting) sink capacity of ears

* LAD = leaf area duration.

The maximum grain yield is thus:
1640 g/m^2/dry weight = 19.1 t/ha (with a 14 per cent water content).
The utilization of solar energy in that case is 4.0 per cent.

In opposition to this calculation the largest rice yields in Japan and the Philippines do not exceed 10 tons (with a water content of 14 per cent). The record yield is thus about half of the calculated maximum yield. Factors responsible for the difference are that if there are no special problems of translocation the yield may be limited by the yield capacity or by the level of assimilate supply. Three cases are possible (Murata 1969): (a) the yield capacity is the limiting factor, the proportion of ripened grains is high, in spite of this an excess of assimilates—the amount that cannot be built in the kernels—remains in the stalk; (b) the supply of assimilates is the limiting factor, the proportion of ripened grains is low; (c) the supply of assimilates and the yield capacity are well balanced, both the percentage and weight of mature kernels are high, only a small amount of assimilates remains in the stalk (Fig. 41).

In category (a) more than 85 per cent of the kernels reach maturity. The yield will be larger with an increase in the number of spikes that can be achieved by nitrogen basic fertilization or topdressing. In the (b) case the setting percentage is lower than 80. In that case the intensity of photosynthesis must be increased by fertilization or CO_2 nutrition. The reason why the (c) variant does not reach the possible maximum is that the intensity of assimilation decreases after heading perhaps due to the "sink-source" feedback interaction.

In connection with these models the advantages of dwarf, semi-dwarf and early types should be emphasized. They are characterized by favourable HI, intensive photosynthesis, and the early ones also by a lower respiration loss.

Further, it should be noted that with changed demands the desirable structure of the model changes too.

In cross-pollinating plant species the ideal type is thought to be represented by the *hybrids* (produced by inbreeding) which will be referred to when analysing the physiological genetics of heterosis.

CHAPTER 7

Heterosis

To explain the phenomenon of heterosis many theories have been set up so far (Bálint 1967, Sváb 1971, Fischer 1978). In the F_1 generation the gene effect may be additive, dominant, overdominant, epistatic etc., that is, the overdevelopment of hybrids may be traced back to various genetic mechanisms.

The quantitative and population-genetic analyses call attention to one thing: there is no genetic system by which all cases can be explained, so it is not possible either that all manifestations of heterosis have the same physiological genetic basis.

There are cases when the hybrid vigour (overdevelopment) can be traced back to a single gene (Stubbe and Pirschle 1940, and others). These cases might encourage us to try to trace the vigour of hybrids to a compound. Considering that it can be hardly supposed that two organisms differ only in a single pair of genes from one another, these investigations must be made with great circumspection.

While not leaving the difficulties out of consideration let us try to review the results of investigations that have tried to explain the phenomenon from the point of view of physiological genetics. These investigations cover 3 fields. They are concerned with differences between the parents and the F_1 generation in the basic metabolic processes (a), in the enzyme systems and hormones regulating them (b); and try to find a relationship between different compounds contained in the plant and seed on the one hand, and overdevelopment on the other.

Physiological differences

Numerous experiments have proved that the hybrids grow more vigorously, give larger yields and accommodate themselves better to the unfavourable conditions of the environment than the parents (Bálint

1967). The reason is to be found in the biologically more efficient system of organic matter production and accumulation. This does not always mean an increased intensity, as e.g. the more vigorous growth of maize hybrids may also be traced back to the higher respiration intensity in the lines compared to the hybrids, beside the higher photosynthetic productivity of the latter.

Photosynthesis

The superiority of hybrids to the lines is manifested in various forms. Their leaf area may be larger than that of the parent lines. Dobrovskaya (in Fyodorov 1968) studying the assimilation surface (LAI) in the double-cross VIR 267 and VIR 37 found that the leaf area per plant of the hybrids exceeded by 83.1 per cent that of the lines.

Fyodorov (1968) compared a single-cross. (V155 × G23 = Krasnodar 3) and a variety line hybrid (Sterling × Krasnodar 3 = Krasnodar 4) for dry weight of leaves in various phases of development (Table 72).

The increase of leaf area can be observed in self-pollinating plant species–e.g. in bean–too. Duarte and Adams (1963, in: Wallace et al. 1972) pointed out a considerable increase in leaf area in field trials and glasshouse experiments alike (Table 73).

However, the larger leaf area of the hybrids is only an advantage when its efficiency is not lower but possibly even higher than that of the parents. Evidence of this is given by the data of Table 74 in which 2 variety hybrids are compared to Krasnodar 3.

The productivity of leaves per unit area is also higher in the hybrids. In the experiments of Dobrovskaya this surplus value was 21.9 per cent. This

Table 72

Leaf area and productivity per m² leaf area of maize hybrids, varieties and lines (Fyodorov 1968)

Hybrid, variety, line	Average leaf area (per plant, cm²)		Productivity (per 1 m² leaf area, g)	
	22 June	12 July	Total yield	Grain yield
Krasnodar 3 hybrid	73.3	367.8	223.0	75.0
V155 line	46.7	304.4	160.9	57.0
G23 line	40.0	198.5	147.0	48.0
Sterling variety	86.5	390.5	132.0	42.0
Liming variety	64.9	422.4	179.0	41.0

difference is based on the higher photosynthetic intensity of hybrids (Heichel and Musgrave 1969, Table 75).

The higher LAD-value of the hybrids also contributes to the total yield, as their leaves remain green for a longer time than the leaves of the lines. Breeding for maturity reached on a green stalk is aimed at increasing this special difference.

In the F_2 generation of hybrids the leaf area is reduced to a minor and the productivity of leaves to a greater extent in comparison with the F_1

Table 73

Genetic analysis of leaf area in field beans (*Phaseolus vulgaris*) (Duarte and Adams 1963)

Genotype	Leaflet number	Per plant basis	
		Leaflet size (cm^2)	Total leaf area (cm^2/plant)
		Field data	
P_1	83	84	6,964
P_2	320	33	10,611
$P_1 \times P_2$ (F_1)	334	58	19,249
		Greenhouse data	
P_1	26.7	147.1	3,936
P_2	63.7	63.9	4,109
$P_1 \times P_2$ (F_1)	68.2	105.5	8,042

Table 74

Leaf area (per plant) in the F_1 and F_2 generations of maize hybrids and its productivity relative to the full grain yield (Fyodorov 1968)

Hybrid	Repro-duction	cm^2 leaf area per plant		For $1m^2$ leaf area		Harvest index (%)
		22 June	12 July	Full crop (g)	Grain yield (g)	
Krasnodar 3	F_1	59.8	353.4	160	57	0.36
Krasnodar 3	F_2	49.3	314.9	132	45	0.35
Pobeda	F_1	61.7	353.7	257	72	0.31
Pobeda	F_2	37.3	343.9	206	63	0.28
Progress	F_1	42.0	272.3	248	36	0.26
Progress	F_2	35.5	244.6	230	31	0.27

Table 75

Differences in net photosynthesis in varieties of *Zea mays* L. (Heichel and Musgrave 1969)

Variety	CO_2 ppm	Leaf temperature (°C)	Photosynthesis (mg CO_2 dm^{-2} h^{-1})	Heterosis (%)
R 181	268	22	40 ± 4	
R 181 × WF 9	251	25	59 ± 3	97
WF 9	271	24	21 ± 2	
WF 9 × R 151	255	22	37 ± 2	48
R 151	268	22	29 ± 3	
WF 9 × OH 43	262	21	31 ± 4	29
OH 43	261	23	27 ± 3	

(Table 74). The same is shown by the decreasing tendency of useful yield as well (Fyodorov 1968).

In the experiment conducted by Hajós-Novák (1978, personal communication) with tetraploid maize hybrids this tendency was not found to be general (Table 76), as suggested by the harvest index values for the leaf area.

The higher productivity of the unit leaf area, on the other hand, is seen in each case.

Sometimes the larger volume of vegetative plant parts is the ultimate goal of heterosis breeding, e.g. with Sudan grass and lucerne hybrids.

By crossing plasmically sterile sweet sorghum with common Sudan grass Barabás (1961, in: Horváth 1963) produced a sweet Sudan sorghum-grass hybrid. The green crop of the hybrid (Hybar Mv 301) exceeded the yield of variety next to it by 31–34 per cent (Table 77).

The hay yield of the hybrid showed a 26–53 per cent surplus in the same experiment. As for the stem/leaf ratio (about 60 per cent) it was more or less equal to the other varieties. Its grain crop exceeded those of the open-pollinating varieties in the experiment by 72 per cent (Horváth 1963).

Böjtös (1976) reports on lucerne hybrids produced by male sterile strains. Of the 12 combinations examined 50 per cent outyielded the control variety Mv Synalfa by 6–12 per cent. On the basis of trials performed at various places it is claimed that it may be possible to attain a 8–10 per cent yield surplus in commercial production. There are data on yield surpluses larger than that in the literature. Considering that the additional cost of seed production consumes a part of the surplus yield (one-third or so), the superiority of hybrids introduced in commercial production must be over 8–10 per cent.

Table 76

Leaf area, grain yield per plant and their heterosis index in the F_1 generations of tetraploid strains and their hybrids in 1978
(Hajós-Novák 1978, personal communication)

Hybrid and strain	Area of leaf attached at the ear (cm^2)	Yield (g)	Productivity per 1 dm^2 leaf area (g)
30 × 35	768.1	170	22.1
35 × 30	869.7	200	23.0
30 (4x)	771.4	160	20.8
35 (4x)	819.6	150	18.3
Heterosis index, %	97	110	113
	109	129	117
30 × 37	840.4	200	23.8
37 × 30	844.6	180	21.3
30 (4x)	771.4	160	20.8
30 (4x)	801.2	130	16.2
Heterosis index, %	107	138	129
	107	124	115
30 × WF-9	822.2	230	28.0
WF-9 × 30 (4x)	838.4	210	25.1
30 (4x)	771.4	160	20.8
WF-9 (4x)	759.9	190	25.0
Heterosis index, %	107	131	122
	110	120	110
35 × 37 (4x)	836.8	200	23.9
37 × 35 (4x)	919.9	210	22.8
37 (4x)	801.2	130	16.2
35 (4x)	815.9	150	18.4
Heterosis index, %	103	143	138
	114	150	132
35 × WF-9 (4x)	808.8	190	23.5
WF-9 × 35 (4x)	867.1	210	24.2
35 (4x)	819.6	150	18.3
WF-9 (4x)	759.9	190	25.0
Heterosis index, %	103	112	108
	110	124	112
37 × WF-9 (4x)	839.0	230	27.4
WF-9 × 37 (4x)	916.1	240	26.2
37 (4x)	801.2	130	16.2
WF-9 (4x)	759.9	190	25.0
Heterosis index, %	107	144	133
	117	150	127

Respiration

We have seen (Heichel and Musgrave 1969) what the average trend of photosynthesis in the maize lines and hybrids was. Göring (1963, in: Bálint 1967) comparing two single-crosses at two different dates pointed

Table 77

Average green yields of variety trials with Sudan grass (Horváth 1963)

Variety	First cutting (q/ha)	First cutting (%)	First and second cutting together (q/ha)	First and second cutting together (%)
	1962 Averages of 6 experimental sites			
Hybar Mv 301	280.4	129.5	381.1	131.0
Krasnodar 1967	196.2	90.6	291.3	100.1
Odesskaya 25	185.7	85.7	285.9	98.3
Kecskeméti édes	189.4	87.5	249.8	85.8
Szegedi 12	231.1	106.7	246.6	84.8
Average	216.6	100.0	290.9	100.0
$SD_{5\%}$	27.7	12.8	62.6	21.5
	1963 Averages of 7 experimental sites			
Hybar Mv 301	291.3	126.1	483.2	132.4
Krasnodar 1967	231.0	100.0	359.4	98.5
Odesskaya 25	224.0	97.0	346.8	95.0
Kecskeméti édes	211.0	91.4	338.6	92.8
Voronezhskaya 1	191.0	85.4	296.6	81.3
Average	230.9	100.0	364.9	100.0
$SD_{5\%}$	22.6	9.8	54.7	15.0

out that the oxygen consumption for g/dry matter was higher in the lines than in the hybrids. Gáspár (1963) studying the respiration of lines and hybrids in 2 mm segments of 3-day-old roots found differences in the lines and hybrids, and even between the individual zones.

Serbanescu (1966) studied the ratio of photosynthesis to respiration in the parent lines and hybrids of the double-cross maize hybrid HD-311, further in variety hybrids of egg-plant, paprika and tobacco. The results are contained in Table 78.

Table 78

Intensity of photosynthesis and respiration (cm³/dm²/h) and trend of chlorophyll content (mg/g dry matter) (Serbanescu 1966)

Plant	Variation	Absolute value of respiration	(%)	Absolute value of photosynthesis	(%)	Absolute value of chlorophyll	(%)
Maize	L-47	2.15	107.5	8.43	83.0	10.8	95.5
	HS ♀	2.00	100.0	10.14	100.0	11.3	100.0
	L-48	2.29	114.5	8.25	81.3	8.8	77.8
	L-49	2.75	141.0	8.48	72.4	11.1	105.6
	HS ♂	1.95	100.0	11.71	100.0	10.5	100.0
	L-50	2.00	102.5	10.17	86.8	10.1	96.2
	HS ♀	2.00	102.0	10.14	116.1	11.3	90.4
	HD-311	1.96	100.0	8.73	100.0	12.5	100.0
	HS ♂	1.95	101.0	11.71	134.1	10.5	84.0
Egg plant	Linia 166	2.02	91.0	31.2	83.6		
	Linia 166 × Lungi	2.20	100.0	37.3	100.0		
	Lungi	1.62	119.0	37.5	100.6		
Paprika	Galben timpuriu	3.70	96.3	34.3	87.0		
	Galben timpuriu × Cecei	3.84	100.0	39.4	100.0		
	Cecei	3.27	85.1	37.1	94.1		
Tobacco	Hicks resistant	2.18	113.5	24.3	106.5		
	Hicks resistant × Linia 128	1.92	100.0	22.8	100.0		
	Linia 128	2.21	115.1	20.6	90.3		

In single-crosses the intensity of photosynthesis is higher than in the parent lines, while the respiration loss lower than the average loss of lines. The photosynthetic intensity is higher in the hybrids than in the parents in the other species too; the loss of respiration of hybrids compared to the parents is lower in tobacco and higher in paprika plants.

Botnar and Soboleva (1974) examined tomato varieties and hybrids for respiration intensity. The results are shown in Table 79 indicating the difference between the crop years as well as the lower respiration intensity of hybrids compared with the parent varieties. The advantage of this is particularly obvious when—similarly to the previously described analyses—the loss of respiration and the intensity of photosynthesis are jointly examined.

The superiority of hybrids is due first of all to differences in the two processes.

Table 79

Intensity of respiration in tomato at various stages of development ($mg\ CO_2/dm^2/h$) (Botnar and Soboleva 1974)

Variety and hybrid	Budding, 25–26 June	Flowering, 6 July	Fruit setting, 11–13 July	Beginning of ripening, 28 July	Ripening, 31 Aug.
1971					
Budanovka	1.51	2.76	2.51	—	1.96
Tamanets × Budanovka F_1	0.24	1.02	0.82	—	0.89
Tamanets	1.51	2.68	2.50	—	2.00
Tamanets × John Baer F_1	0.34	1.38	0.93	—	0.72
John Baer	0.82	1.92	1.49	—	1.38
1972					
Budanovka	0.21	0.73	0.35	0.17	—
Tamanets × Budanovka F_1	0.14	0.72	0.28	0.11	—
Tamanets	0.48	0.83	0.69	0.74	—
Tamanets × Stalovyi 208	0.42	0.76	0.61	0.50	—
Stalovyi 208	0.53	0.86	0.66	0.69	—

Mineral nutrition

As seen in the chapter "Nutrient utilization" genetic differences can be pointed out in this respect too, and the nutritive elements have an indirect effect on the processes of building up and decomposition.

The nitrogen status of the plant has an influence on both the assimilation of the organism and the uptake of many elements (Nelson 1966). Thus it is important to get information about the nitrogen utilization of hybrids. In the course of examinations attention should be paid to the fact that maize for example does not utilize the different inorganic nitrogen compounds in the same way. The young maize plant makes better use of ammonia, while the fully developed plant of the nitrate nitrogen sources. Harvey (1939) studied the phenomenon in inbred maize lines and their hybrids, further in tomato hybrids and parent strains, and pointed out genetic differences in the uptake of nitrate and ammonium nitrogen, and of potassium.

Burkholder and McVeigh (1940, in: Bálint 1967) compared three single cross hybrids and their lines supplied with increasing rates of nitrogen fertilizer and found 15, 68.5 and 10 per cent surplus nitrogen incorporation with the lower and 192, 191 and 105 per cent with the higher nitrogen concentration, calculated for fresh and dry weight. The experiment does not give any information as to whether it was the lines or the hybrids that used less nitrogen for the production of a unit amount of dry matter.

Holbert and De Turk (1938, in: Bálint 1967) examined the organic matter production and protein percentage in low (L) and high (H) protein content Illinois lines. At the age of 88 days the two lines grown in water culture had the same dry weight with a 25 ppm N supply; with 100 ppm nitrogen the H-line produced 42.5 per cent more dry matter than the L-line. At the same time, the protein percentage of the H-line was higher in both cases. This indicates that the H-line incorporates more nitrogen not only in the kernels but also in the leaves. As for the conditions of nitrogen uptake and utilization no data have been supplied by the authors.

In Fyodorov's (1968) experiments a comparison between maize lines and their hybrids reveal that the amount of nitrogen incorporated in various organs was larger in the hybrids than in the lines. There are no data on nitrogen utilization relative to dry matter production.

Of the mineral nutrients phosphorus has been most frequently studied in maize for some decades. Investigations into the phosphorus utilization of plants were already made in 1926. Hoffer (1926) studied the ash components in a single cross hybrid and its parent lines. In the ashes of young plants raised in glasshouse on clay soil 1.09, 0.95 and 0.91 per cent phosphorus was detected for the A- and B-lines and A × B hybrid, respectively. When grown in loam the plants of line B used about 25 per cent less phosphorus than the hybrid plants for building up a unit amount of dry matter.

In experiments carried out some ten year later Smith (1934) found the maize hybrids to be decidedly superior to the inbred lines as regards phosphorus uptake. The difference was especially remarkable in the case of a poor phosphorus supply. This advantage of the hybrids is partly attributed to the more developed root system. Furthermore, when crossing inbred strains of low phosphorus utilization the author failed in producing high phosphorus utilization hybrids.

In the progeny of inbred strains with low and high P-utilization, respectively, Lynese (1936) observed an intensive phosphorus uptake which suggests the dominant inheritance of this property. In the course of studying the radioactive phosphorus uptake in inbred lines and hybrids Rabideau et al. (1950) found that the hybrids absorbed more radioactive phosphorus than their parents. This can be explained by the more rapid initial development and increased metabolism of hybrids. In the examined combination (A × B) the bulk of roots and phosphorus uptake of the hybrid—in agreement with Smith's findings—exceeded those of the parent lines, though as regards the ratio of shoot to root the superiority of the hybrid was not observed in every case.

Gáspár (1963) studied the relationship between hybrid vigour and phosphorus uptake in maize seedlings (Mv 5 hybrid, A, B and A × B strains) raised in water culture up to the age of 2 weeks. With the equipment constructed by himself the author was able to follow the course of P^{32} uptake. On a fresh weight basis the root/leaf ratio was not more favourable in the hybrid than in the A strain, but with the average of the parent strains taken for 100 the hybrid took up 25 per cent more phosphorus, and compared to the better strain this surplus was 29 per cent. On the basis of these data the author–in agreement with Whaley (1952)—considered the higher phosphorus uptake to be one of the causes of hybrid vigour. In these experiments P was not regarded as a nutritive element but as the index of the intensity of metabolic processes. Namely, it is only on the 22. to 25. day after emerging that the plant begins to use the phosphorus supplies of the soil for the purpose of nutrition.

There are no data on whether differences in P-uptake have been made use of in the work of breeding.

Of the *microelements* boron was studied by Sayre (1955) in the leaves of 18 Ohio lines; the author pointed out differences in the percentage of boron between the lines.

Radov et al. (in: Bálint 1977) studied the occurrence of microelements in the grains of various hybrids. Some of the hybrids examined did not contain all microelements (e.g. zinc, strontium etc.). The following hybrids and varieties had the fullest range: VIR 42, Dnepropetrovsk 3, North-Dakota, Harkov 23.

The breeding value of these investigations lies in the fact that they point out the differences in microelement content between the individual lines and hybrids. On the basis of diallele crossings Thomas (1963) established the additive character of genes determining the inheritance. No information has been supplied, on the other hand, on the relationship between microelement content and organic matter production in lines with different accumulative capacities. Studies on this aspect and on the inheritance of microelement utilization may provide a basis for producing hybrids with a lower demand for microelements or higher efficiency in utilizing them.

Table 80

Microelement content in maize lines and their hybrids* (Bálint 1967)

Lines, single crosses	Mn		Cu		Mo	
	ppm (dry ♀ matter) (g)	(%)	ppm (dry ♀ matter) (g)	(%)	ppm (dry ♀ matter) (g)	(%)
C5	130	96	6.20	101	0.65	125
C5 × 014	135	100	6.10	100	0.52	100
014	155	115	5.66	93	0.47	94
WF9	110	100	5.54	105	0.22	73
WF9 × M14	110	100	5.20	100	0.30	100
M14	110	100	7.10	136	0.52	173
T18	115	100	4.47	106	0.40	143
T18 × WF9	115	100	4.18	100	0.28	100
WF9	110	104	5.45	130	0.22	79
T30	105	87	7.14	103	0.49	84
T30 × T31	120	100	6.90	100	0.58	100
T31	110	91	6.50	94	0.62	107
0118b	125	104	7.12	103	0.64	110
0118b × M14	120	100	6.90	100	0.58	100
M14	110	91	7.10	103	0.52	89

* For the sake of making comparison easier, when calculating the percentages the values were established by taking the hybrid for 100.

In experiments including 5 single-cross hybrids (Bálint 1967) three microelements manganese, copper and molybdenum were studied. The results are summed up in Table 80.

A comparison between the data of hybrids and those of parents supplies rather diverse results. In 11 of 15 cases an intermediate microelement level is found. When compared to the maternal parent the hybrids are superior in 4 while more or less inferior in 9 cases.

As for the individual microelements, molybdenum in each case shows an intermediate heredity while manganese with the exception of a single case, often characteristic of the inheritance of quantitative features. With copper, on the other hand, a setback relative to the parents' average was observed in three cases.

The data are encouraging in our study of the role and distribution of microelements. It is also necessary to follow their accumulation under conditions of severe microelement deficiency, at different rates of basic fertilization on various soil types.

By the interpretation of the diversified data series two correlations can be said to be general. One of them is that the root development of the hybrid exceeds in every case that of the parents (Table 81, Paddick 1944, in: Suresh K. Sinha and Renu Khanna 1975).

Heterosis indicated by the root weight does not involve an improvement in the shoot/root ratio.

The situation is similar with the uptake of nitrogen, phosphorus and potash (Table 82, Rao and Venkateshwarlu 1971, in: Suresh K. Sinha and Renu Khanna 1975).

The hybrid in each case took up more nutrients than the parent lines. No essential differences in the proportions of nutritive elements were found between the organs.

Investigations these days, are aimed at trying to find out how much the lines improve the nutrient reaction of the hybrid and influence the total

Table 81

Shoot : root ratio in maize inbreds and hybrids (Paddick 1944, in: Bálint 1977)

Strain	Root dry weight (mg)	Shoot dry weight (mg)	Shoot : root ratio
La	194	247	1.27
Pr	161	268	1.66
La × Pr	327	428	1.30
Pr × La	385	477	1.23

Table 82

Distribution of nutrients per plant and on unit weight basis in a sorghum hybrid (CSH 2) and its parents (Rao and Venkateshwarlu 1971, in: Suresh K. Sinha and Renu Khanna 1975)

Nutrient	Leaf			Stem		
	P_1	P_2	$P_1 \times P_2$ (F_1)	P_1	P_2	$P_1 \times P_2$ (F_1)
	Milligrams per plant					
Nitrogen	254	251	381	127	150	232
Phosphorus	26.7	26.6	37.4	17.1	18.2	26.6
Potash	125.7	148.3	187.3	216.7	229.8	374.7
	Percent in plant tissue					
Nitrogen	3.10	2.96	3.15	2.34	2.57	2.40
Phosphorus	0.33	0.31	0.31	0.32	0.31	0.27
Potash	0.53	1.75	1.55	4.00	3.94	3.89

yield and the protein content of grain (Powell 1968, Pollmer et al. 1979) rather than determining the extent of the hybrid's superiority to the lines in nitrogen utilization.

The economy of water utilization in maize hybrids was first studied by Kiesselbach (1926) who determined the amount of water required for the production of 1 g dry matter and 1 g ear crop in 3 single-cross and 2 double-cross hybrids composed of 6 lines. This amount was in the above order 319.0, 260.0 and 254 g, respectively. The lines were obtained from the open pollinating variety Nebraska White Prize, in which the corresponding value was 242.7 g, that is more favourable than in the hybrids. The hybrids used about 20 per cent less water than the lines for 1 g dry matter production. In respect of the amount of water used for 1 g grain yield the differences were much greater. Compared to the 900.0 g value of the lines, the hybrids only required half of that amount (443.0 and 441.0 g, respectively). In this respect the value of the hybrids is slightly better than that of the initial variety (469.9).

On the basis of a subsequent investigation (Kiesselbach et al. 1935, in: Bálint 1967) the superiority of hybrids was partly explained by the fact that in the F_1 generation the average length of major roots exceeded by 45 per cent and the maximum depth of their penetration by 55 per cent the respective values of the parent lines. According to Maximov and Kohanovskaya (1921) the transpiration coefficient of maize considerably varies from year to year. In 1911–17 at the Agricultural Research Institute of Bezenchuk, the limits of this variation were 144 and 439 g water/g dry matter. The nutrient status also has an influence on it. However, results obtained with different varieties and hybrids suggest—in agreement with the author's own investigations—that this property of the plants can be improved by selection (and by crossing the selected lines).

All this discussion on certain elements of the metabolic processes raises the idea that all these properties ought to be studied in the same combinations as their parent forms, so as to acquire a wider knowledge than at present.

Development of hybrids

In the course of growth and development the larger leaf area results—naturally—in a larger dry matter yield, and occasionally even in a larger grain yield, as shown by the example of two wheat hybrids (Lupton 1976).

The hybrids—with a single exception (the Maris Ranger × Primepi combination in 1972)—were always found to be superior in productivity to the parents' average; in comparison to the better parent, however, the situation was not as favourable as that (Table 83).

These moderate positive differences are reflected in the growth rate and leaf area index as well (Fig. 43).

On the basis of analysing the intensity of photosynthesis the results for the combination Maris Ranger × TW 74 are interpreted as seen in Table 84.

Intensive heterosis effect is felt in the intensity of photosynthesis 12 days before flowering. In that phase the hybrids grow more vigorously than the parents. The hybrid is also more responsive to increasing light intensities. Later the photosynthetic intensity of the hybrid falls back between the values of the two parents.

Table 83

Grain yields for 1971 (a) and 1972 (b) from F_1 growth analysis experiments (g/m^2) (Lupton 1976)

Parents	TW 74				Primepi	
	a	b	a	b	a	b
			795	739	577	1,075
TB 306/45	600	789	947	934	867	1,001
Maris Ranger	782	856	936	850	785	797

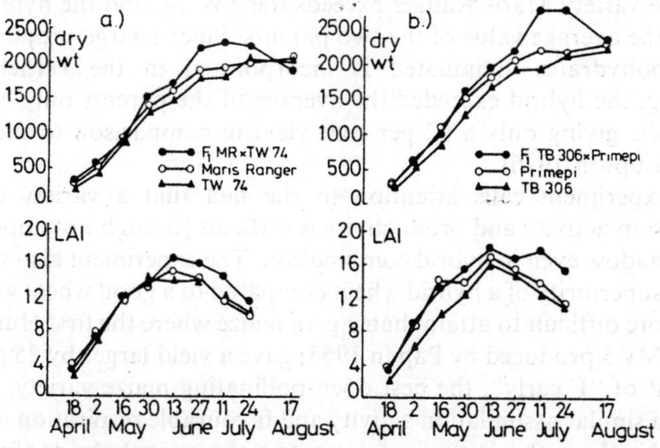

Fig. 43. Growth analysis 1972: leaf area index (LAI) and dry weight development of F_1 hybrids and parent varieties of Maris Ranger × TW 74 and TB 306 × Primepi (Lupton 1976)

Table 84

Rate of photosynthesis (mg CO_2 dm^{-2} h^{-1}) measured under field conditions at 30 cal cm^{-2} h^{-1}; linear regression coefficient of rate of photosynthesis on light intensity; and percentage of assimilated carbon recovered in the grain at maturity (Lupton 1976)

	Anthesis − 12 days	Anthesis	Anthesis + 14 days	Anthesis + 28 days
Maris Ranger				
Rate of photosynthesis	17.80	17.20	19.60	8.70
Regression coefficient	0.59	0.41	0.44	0.18
Percentage carbon recovered in the grain	1.50	8.80	67.30	75.70
TW 74				
Rate of photosynthesis	17.40	18.30	13.40	11.10
Regression coefficient	0.55	0.58	0.34	0.26
Percentage carbon recovered in the grain	2.80	8.00	36.50	61.30
F_1 Maris Ranger × TW 74				
Rate of photosynthesis	19.40	18.10	16.50	9.30
Regression coefficient	0.64	0.57	0.34	0.13
Percentage carbon recovered in the grain	6.30	11.00	48.10	66.20

The extent of ^{14}C incorporation in the grain is also greater in the first period. In that phase there are hardly any difference between the varieties. Later the variety Maris Ranger exceeds the TW 74, and the hybrid only reaches the average value of the two parents. Since a large proportion of the carbohydrates assimilated is incorporated in the kernels after flowering, the hybrid exceeded the average of the parents only by 2 per cent, while giving only a 92 per cent yield in comparison to the better parent (Lupton 1976).

The experiment calls attention to the fact that a variety of high assimilation activity and productivity is difficult (though not impossible) to overshadow even in hybrid combination. The experiment also suggests that the superiority of a hybrid wheat compared to a good wheat variety is much more difficult to attain than e.g. in maize where the first Hungarian hybrid (Mv 5 produced by Pap in 1953) gave a yield larger by 25 per cent than that of "F early", the best open-pollinating maize variety.

With a similar assimilation activity and favourable respiration rate—as observed in the author's own experiments with variety hybrids of maize—the higher harvest index, i.e. the more favourable accumulation of assimilates, may result in the superiority of hybrids. In experiments carried out by Dogget (in: Suresh K. Sinha and Renu Khanna 1975) with grain sorghum the harvest index was 0.32 for one of the parents (P_1) and 0.37 for the other (P_2), and the hybrid (F_1 generation of $P_1 \times P_2$) reached a value of 0.41.

Enzymes and hormones

The intensive growth of hybrids, their increased assimilation activity etc. have given rise to the idea that a compound of high biological efficiency, due to increased metabolic intensity has to be found. Some of them are shown below.

The intensive growth of hybrids called attention to the growth substances present in the seeds. Experiments in this field began in the early forties.

However, no relationship was found between the vegetative vigour of the hybrids and the amount of auxin stored in the seed. In spite of the lower growth intensity displayed by them varieties of sugary endosperm contained a higher percentage of auxin than those with a waxy endosperm.

A comparison of vigorously growing tetraploid to diploid forms could not detect any definite relation between auxin content and growth intensity either.

In the author's experiments analysing the kernels in the Mv 5 and Mv 1 maize hybrids and their parent forms no relationship between the auxin content and the superiority of the hybrid was found again.

After the unsuccessful attempts made with auxin and growth substances belonging to the B-group the attention has turned towards the enzymes; most experiments are at present conducted in this field.

It is to the hybrid character of enzyme proteins found in the hybrids that Schwartz (1962) traced the superiority of crossed forms. According to his data the E esterase gene in the maize tissues is determined by two alleles. The author separated various E-type genes—E^F, E^N, E^S—on the basis of electrophoretic mobility. Their hybrid combinations—E^F/E^N, E^F/E^S, E^H/E^S—showed an intermediate migration rate between the two alleles. From this the author drew the conclusion that as a result of crossing, hybrid enzymes were formed. He even constructed a model for how the RNA molecule controlled the synthesis of the enzyme. The intermediate level is characteristic of a number of physiological parameters. Between the esterase enzymes and the hybrid superiority no linear correlation could be detected.

Table 85

Arithmetic averages of enzyme activities in maize lines and hybrids expressed in hundredfold extinction value (E × 100), and the smallest significant differences (SD) (Marsálek 1972)

Genotypes	Year of examination				
	1966	1967	1968	1969	Averages
			Polyphenoloxidase		
A 166	29.54	29.34	24.40	20.73	26.00
A 188	29.17	33.81	26.68	21.79	27.86
A 34	29.28	28.51	24.69	20.62	25.76
A 171	28.97	27.68	35.61	31.00	28.32
(A 166 × A 188)	27.86	26.63	25.10	20.86	25.11
(A 34 × A 171)	26.27	26.85	26.61	21.26	25.25
SD = 5%	1.12	2.35	1.53	1.21	1.55
SD = 1%	3.24	1.09	2.15	1.70	2.05
			Peroxidase		
A 166	60.83	49.41	45.37	59.65	53.82
A 188	56.51	61.45	47.88	63.01	57.21
A 34	55.76	57.62	46.50	52.52	53.10
A 171	56.31	55.87	61.42	59.26	58.22
(A 166 × A 188)	61.20	60.57	47.08	61.44	57.57
(A 34 × A 171)	49.02	56.37	47.05	50.96	50.85
SD = 1%	1.39	1.38	1.44	1.62	1.46
SD = 5%	1.96	1.93	2.02	2.27	2.05
			Ascorbateoxidase		
A 166	2.50	0.98	4.82	1.78	2.52
A 188	2.59	0.98	3.48	1.78	2.21
A 34	3.85	2.45	4.10	4.18	3.65
A 171	4.45	2.14	5.34	2.05	3.49
(A 166 × A 188)	1.85	1.12	4.73	1.54	2.31
(A 34 × A 171)	5.51	1.58	7.24	3.19	4.38
SD = 5%	1.00	0.45	0.73	0.88	0.77
SD = 1%	1.40	0.62	1.02	1.24	1.07

Marsálek (1972) studied the enzyme activity of polyphenoloxidase, peroxidase and ascorbate-oxidase in 2 single-cross hybrids and their parent strains. In the experiment conducted for four years measurements were taken every year at the 1-, 2- and 3-leaf stages. The results of measuring (averaged by year) are shown in Table 85.

Table 86

Activity of succinate-dehydrogenase in the two-way maize hybrids Svetoch and Slava and in the initial forms (Yakovlev and Raskova 1975)

Variant	Svetoch (40 × 43)	VIR 40	VIR 43	Slava (44 × 38)	VIR 44	VIR 38
			For 1g fresh weight			
A	548 ± 65	482 ± 32	266 ± 32	434 ± 14	474 ± 15	252 ± 13
B	128 ± 44	138 ± 34	0	48 ± 7	116 ± 19	26 ± 4
V	293 ± 44	259 ± 29	105 ± 22	287 ± 40	322 ± 36	224 ± 14
			For 100 germ			
A	4,488 ± 318	3,104 ± 245	1,743 ± 173	2,760 ± 21	3,357 ± 113	1,855 ± 140
B	1,624 ± 297	1,110 ± 207	0	398 ± 36	672 ± 97	189 ± 29
V	2,472 ± 328	1,660 ± 175	678 ± 135	1,299 ± 226	2,444 ± 226	1,354 ± 115

Note: Succinate-dehydrogenase activity: in the presence of 1.66×10^{-2} M succinate (A), 1.66×10^{-2} succinate and 1.66×10^{-2} M malonate (B), 8.33×10^{-2} M succinate and 1.66×10^{-3} M malonate (V).

The tendencies observed in the data suggest that there were genetically determined differences, but the mode of response was not uniform in the two combinations.

Yakovlev and Raskova (1975) examined the activity of succinate-dehydrogenase, an enzyme of importance in the energetic system of the cell, in seedlings of 2 single-cross hybrids and of the lines forming them. The results are summed up in Table 86.

The better parent is surpassed by one of the hybrids (Svetoch) but not by the other. The situation is similar in the case of a double-cross combination produced from the 2 single-crosses.

The experiments described may be criticized for the small number of combinations included in them. In diallele crossing series of 10 lines, Hageman et al. (1974) tried to find out which of the lines had a good general combining ability for the level of NRA. The results in Table 87 indicate that the level of NRA was increased in all combinations by two lines (line 2 and 8). The increase affected only the level of nitrate-reductase activity and not the total yield. From the point of view of total yield the combinations 2 × 7, 5 × 9, 3 × 9 and 1 × 4 were the best. The highest protein output was found in the 1 × 4 combination (905 kg/ha). As for the level of NRA line 1 was placed second and line 4 tenth.

Analyses of relations between the enzyme levels and the organic matter production are inadequate if only because the individual characters were mostly studied on average samples. Fadeeva et al. (1975) studied the peroxidase isoenzyme content of radish varieties and lines on individual and average samples alike. The frequency of occurrence of the 9 types—B_1–B_9—found in the Virovsky variety in 74 plants was the following (the figures in brackets represent the number of occurrence): B_1(26), B_2(12), B_3(15), B_4(3), B_5(2), B_6(7), B_8(2), B_9(2). The morphological balance—as seen—means no biochemical homogeneity.

Table 87

Relative nitrate reductase activity, grain yield and grain protein for 10 inbreds and a partial diallel of 45 hybrids. Data arranged in order to decrease nitrate reductase activity (NRA) (Hageman et al. 1974)

Genotype	Relative NRA	Yield, bu A^{-1}	Protein, kg ha^{-1}	Genotype	Relative NRA	Yield, bu A^{-1}	Protein, kg ha^{-1}
2	22.4	61	451	1×3	19.1	131	823
1	19.2	69	515	3×9	18.7	138	790
9	18.0	84	557	7×8	18.5	80	500
3	16.9	99	638	3×5	18.2	131	833
8	15.8	92	701	9×10	18.2	113	697
7	14.9	90	614	5×8	17.6	111	684
6	14.5	84	588	3×6	17.2	103	643
10	11.4	70	593	3×7	17.0	122	728
5	10.9	95	663	2×5	16.8	100	680
4	9.4	87	649	1×6	16.8	112	755
Avg	15.0	83	601	1×4	16.7	137	905
2×8	27.9	106	687	4×8	16.7	113	707
2×7	27.3	139	844	3×4	16.6	129	792
1×8	25.8	118	779	1×10	16.4	110	757
1×2	23.6	124	798	7×10	15.8	102	701
2×9	23.5	118	709	3×10	15.7	100	664
1×9	22.9	130	738	6×9	15.7	111	692
3×8	22.6	122	736	4×9	14.4	120	700
8×10	22.1	104	698	5×10	14.0	105	703
2×10	21.9	102	691	6×7	13.8	92	575
2×3	21.8	118	756	5×7	13.1	115	754
2×4	21.0	126	836	5×6	12.6	87	564
2×6	20.8	103	667	4×10	12.1	100	663
8×9	20.6	97	594	4×7	12.0	128	797
1×7	20.2	135	868	6×10	11.5	98	640
7×9	19.9	121	725	5×5	10.6	95	663
1×5	19.6	119	812	4×6	10.6	101	640
6×8	19.6	90	601	Avg	18.1	113	720
5×9	19.2	137	812				

Table 88

Alleles observed in the eight strains of corn studied (Heidrich-Sobrinho and Cordeiro 1975)

Tissue / Enzyme / Loci / Strain	Plumula esterase		Roots		Plumula per-oxidase	Ac. phosph.	Endosperm catalase	Alcohol dehyd.
	E_1	E_3	E_{11}	E_4	Px_{11}	Ap_1	Cat_1	Adh_2
Tuxpan 94.6224	F	S	F	C	S	S	F	F
Tuxpan 1020.17	F	S	F	C	S	S	S	F
T 61.984–1	F	S	F	F	S	F	F	F
NC 83.7628	S	S	S	C	S	S	S	S
FB 4. 5941	S	F	F	E	S	F	S	F
Salbert 5392	F	F	F	E	F	S	S	F
SR 201.5232	S	F	F	D	F	F	S	F
SR 527.5260	F	F	F	D	F	S	F	F

Note: S = slow allele; F = fast allele; C, D, E, F correspond to the four alleles of E_4 according to Harris nomenclature, from the slower C to the faster F allele.

According to the results of enzyme analyses the cumulative co-dominance of the enzyme systems of parents (they jointly control the development of the progeny) is the basis of heterosis.

Beside the diversity of alleles which jointly increase the intensity of metabolic processes in the progeny the activity level of iso-enzymes should not be underestimated.

Having determined the enzyme patterns of 8 maize strains (Table 88) Heidrich-Sobrinho and Cordeiro (1975) found a relationship between enzymatic difference index and general combining ability (Table 89).

This difference is shown in the fact that in NC 83 and other strains of high combining ability there is a multi-allele while in the poor combining ability Tuxpan 94,6224 only a single-allele difference.

Fedak and Rajhathy (1972) studying the activity of esterase and amylase iso-enzymes in barley hybrids found a total additive inheritance combined in some cases with epistatic and overdominant gene effects. In the hybrids examined the iso-enzymes and the hybrid vigour seem to be jointly inherited.

Relying on the enzyme experiments Lorenz (1975) tried to make selection for combining ability on the basis of the amount of free amino acids forming the product and substrate of enzymatic responses. For the correlation between the ratio of asparagin accumulation to the amount of free amino acids in older leaves and the volume of grain yield he obtained a

Table 89

Average yields (kg/ha) of the specific and general combinations and the enzymatic diversity index (Heidrich-Sobrinho and Cordeiro 1975)

	Tuxpan 1020	T 61.984–1	NC 83.7628	FB 4.5941	Salbert 5392	SR 201.5232	SR 527.5269	General combining ability
Tuxpan 94.6224	2,887 (1)	5,380 (2)	7,040 (4)	5,078 (5)	6,720 (4)	5,401 (6)	5,741 (3)	5,607 (25)
Tuxpan 1020.17	×	5,360 (3)	5,861 (3)	3,667 (4)	6,704 (3)	5,807 (5)	4,913 (4)	5,028 (23)
T 61.984–1		×	6,067 (6)	6,347 (4)	6,201 (5)	6,293 (5)	6,496 (4)	6,163 (29)
NC 83.7628			×	6,916 (5)	6,173 (6)	6,473 (6)	6,627 (7)	6,451 (37)
FB 4.5941				×	6,353 (3)	6,433 (2)	6,207 (5)	3,857 (28)
Salbert 5392					×	5,153 (3)	7,067 (2)	6,339 (26)
SR 201.5232						×	6,008 (3)	5,938 (30)
SR 527.5260							×	6,151 (28)

Table 90

RNA content (mg/100 seedlings) of seedlings in maize lines, varieties and hybrids (Gilyazetdinov et al., in: Konarev 1971)

Lines and hybrids	RNA content of		RNA/DNA ratio in		Increase of RNA in 7 days
	1-day old	8-day old	1-day old	8-day old	
	seedlings				
VIR 44	5.61	52.50	9.1	5.3	9.4
VIR 38	5.74	57.20	8.7	8.2	10.0
Slava F_1	6.01	70.50	8.8	5.5	11.7
VIR 40	5.49	60.50	8.7	5.0	11.0
VIR 43	5.74	52.50	8.7	3.9	9.1
Svetoch F_1	4.86	57.00	8.8	6.3	11.7
VIR 42 F_1	5.22	64.00	6.7	4.7	12.3
Voronezhskaya 76	5.05	49.50	4.4	2.8	9.8
Odesskaya 27 F_1	6.22	71.22	9.4	3.5	11.5
Gloriya Janeckava	7.36	55.94	9.1	4.8	7.6
Harkovskaya 44	8.35	33.05	8.4	4.3	4.0
Bukovinskii 3 F_1	7.93	68.64	8.5	2.9	8.7

value of $r = -0.76$ in culture fluid and $r = 0.81$ in the field. This index is related to the accumulation and transportation of organic nitrogen. Sarcosine (N methyl glycine/glycine + serine ratio) when determined in young leaves provides a basis for forecasting the yield ($r = -0.74$). For practical use methods suitable for serial examinations must be evolved.

To complete the picture we have to go back to the genetic code that controls the synthesis of proteins and enzymes. Of the RNA, RNA/DNA ratio and rate of RNA synthesis in maize seedlings a survey is given in Table 90 (Gilyazetdinov et al., in: Konarev 1971).

Slava F_1 is superior to its parents, but in the case of Svetoch F_1 the situation is different. Therefore the ratio of bases forming the nucleic acids is considered to be more characteristic than the total amount of DNA and RNA. Belozerskii (1959) determined the basic structure of DNA and RNA for various plant species. After his work Semenenko (1964) determined the ratio of bases forming the RNA in germinating seeds of maize lines and single crosses. Data on the parent strains of the hybrid VIR 42 are contained in Table 91.

No essential differences between the hybrids and lines can be demonstrated. The difference in purine/pyrimidine ratio is not substantial either.

Table 91

Nucleotide composition of RNA in maize lines and -hybrids (Semenenko 1964)

Lines and hybrids	Number of elementary particles	Nucleotide content in mole %				Purine b / Pyrimidine b	G+C / A+U
		A	G	C	U		
		M±m	M±m	M±m	M±m	M±m	M±m
26	3	22.8±0.78	27.9±1.57	26.0±1.57	24.4±1.39	0.90±0.08	1.12±0.08
27	5	20.8±1.27	34.7±0.87	20.7±0.98	28.3±1.70	1.13±0.13	1.17±0.04
26×27	7	22.9±1.14	28.6±1.64	24.4±0.95	25.9±1.54	1.02±0.10	1.08±0.05
28	7	21.0±1.15	28.1±0.87	23.8±0.87	25.2±0.85	1.00±0.08	1.12±0.07
29	7	20.2±1.13	26.2±1.13	26.0±0.77	23.2±1.40	0.94±0.08	1.20±0.05
28×29	8	22.4±1.14	28.3±0.64	26.8±0.67	23.5±1.21	1.00±0.08	1.20±0.10

At the end of the chapter on heterosis reference was made to the necessity of a complete examination of metabolism. The above discussed analyses ought to be connected with the basic metabolic test.

Prognosis

The most difficult part of heterosis breeding is the determination of combining ability in varieties and lines. Among a thousand newly produced maize lines there is usually not more than one line that gives a better combination than the existing ones. The breeder's work would be greatly facilitated if he were able to make a prognosis from analysing the grain yield of lines to be crossed. It would be similarly advantageous if the superiority of the hybrid could be foreseen from the examination of seedlings, or the combining ability estimated from the analysis of the leaves of lines.

It would be easier to solve the problem if the combining ability were determined by a single gene and the biochemical mechanism of this process known. Li and Rédei (1969) having crossed *Arabidopsis* mutants blocked at various stages of thiamine synthesis found a considerable superiority in the heteroallele hybrids, based on the dominant complementary effect of the two unconnected alleles. In some cases the complementation can be interpreted by superdominance.

In polygenically determined systems prognoses are more difficult to make, though the first attempts were—in this case too—based on the more complex "richer" biochemical system of hybrids.

Robbins (1941) studied the influence of an extract of inbred and hybrid maize grains on the growth of a *Phycomyces Blakesleeanus* culture (Fig. 44). In this figure the extract made from maize grains (1 ml/l grain) was added to a culture medium containing sugar, mineral salts, thiamine and asparagine. A = 4–8 lines, B = 187 lines, C = 4–8 × 187, D = 187 × 4–8.

He attributed the effect of maize added to a culture fluid containing sugar, asparagine, mineral salts and thiamine to special growth substances and called it Z-factor.

The amount of the Z-factor is larger in the endosperm than in the embryo, and larger in the hybrid grain than in the parents. The biotin content of the extract is related to the Z-factor. The Z-factor was broken down to a Z_1 fraction absorbed by active carbon and to a non-absorbable Z_2 fraction. Z_1 was identified as hypoxanthine, Z_2 may be a mixture of amino acids.

These analyses were precursors to those experiments by which the heterosis was to be explained with the presence and ratio of growth substances in the leaf and seed.

Matzkov and Manzyuk (1962) considered the superiority of hybrids to be due to growth substances contained in the leaves rather than to those

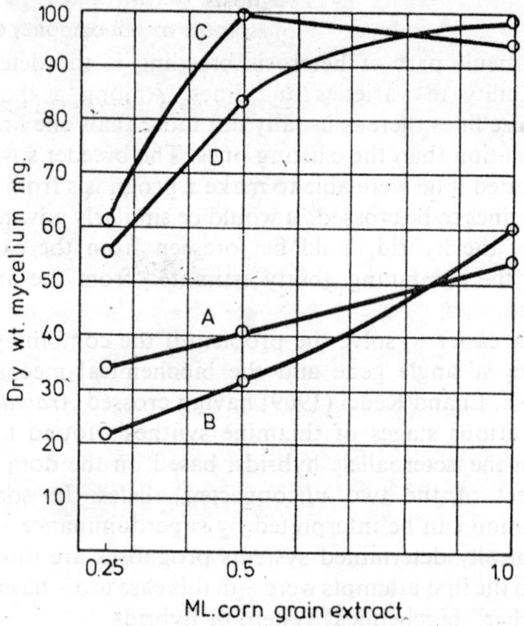

Fig. 44. Increase in the dry weight of Phycomyces produced by extracts of air dry grains maize (Robbins 1941)

present in the seed. These substances set the foundations for the superiority of hybrids in growth and development by increasing the leaf area and thereby the organic matter production. Since the size of the cell generally is not larger in the hybrid than in the parents, the larger leaf area is due to the higher number of cells; here the vitamins of leaves play an important role: biotin, B_1, B_2, B_6, B_3, B_c, and the nicotine acids.

To determine the amount of growth substances belonging to group B *Saccharomyces cerevisiae* cultures were used. The suspension of the fungus was placed in a culture fluid mixed with the extract of the leaves of the plant. The values of growth obtained with the different extracts are shown in Fig. 45.

In each triade the left-hand column represents the maternal, the right-hand one the paternal form, and the middle column refers to the hybrids. The overdevelopment of hybrids is explained by the more complex stock of substances belonging to group B. The less complete series of growth substances contained in the lines are completed in the hybrid.

The method of Matzkov and Manzyuk (1962) was tested by Kovács (1972) on 6 inbreds and variety hybrids; she added an extract made from grains instead of leaves to the *Saccharomyces cerevisiae* culture. In the hybrids produced from the lines the overdevelopment of hybrids corresponded to the growth of cultures; in one of the varieties and in a variety hybrid, on the other hand, linear correlation was not observed.

After the analysis of 200 combinations Gáspár (1964, personal communication) could not point out a linear correlation.

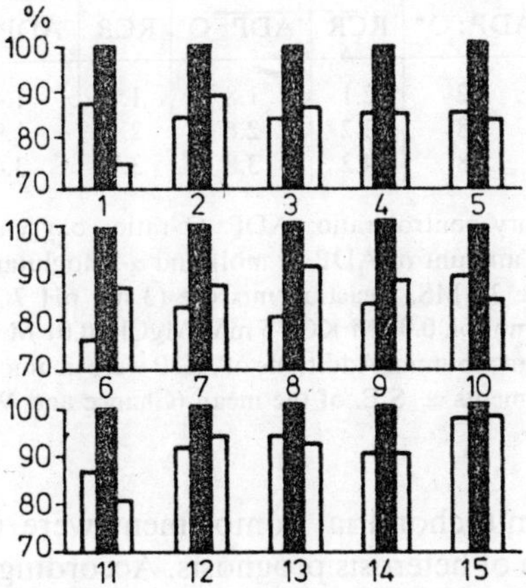

Fig. 45. **Growth substance belonging to group B in maize hybrids and their parent forms as pointed out by yeast-test, with the data of hybrids taken for 100% (Matzkov and Manzyuk 1962)**

Manzyuk (1973), a collaborator of Matzkov, who developed the method further, analysed the growth stimulation of vitamins of group B in the lines (nicotine acid, biotin, inositol) and of the mixture of grains. He placed the lines in groups of good and poor combining ability, respectively, on the basis of their vitamin contents. A mixture of extracts from the grains of various lines resulted in every case in a 3–51 per cent growth in yeast cultures, but of the 151 hybrids 28 per cent showed poor combining ability. The hybrids of high combining ability lines while showing a 51 per cent increase in yeast culture displayed only an 11 per cent surplus productivity. The prognosis was confirmed to 66 per cent in 1966 and 83 per cent in 1967. The amount of inositol was higher in the poor combination ability lines than in the good combinations. Therefore the method is thought to be suitable for forecasting the performance of sublines produced from the same line.

Table 92

Respiration of mitochondria of two lines of wheat and their hybrid. Concentration of substrates was α-ketoglutarate 10 mM, succinate 8 mM, NADH 2 mM (Sarkissian 1972)

Source of mitochondria	α-ketoglutarate		Malate		Succinate		NADH	
	RCR*	ADP : O*	RCR	ADP : O	RCR	ADP : O	RCR	ADP : O
28	3.7	3.2	2.1	1.8	1.7	1.4	4.7	1.3
31 MS	4.6	3.8	3.2	2.6	2.1	1.9	5.2	2.0
31 MS/28	5.5	5.8	4.2	3.8	3.7	3.4	7.0	2.4

* RC ratio: respiratory control ratio; ADP : O ratio: oxygen utilized (μ atoms O_2) responding to a known amount of ADP (μ mol) and α-ketoglutarate as substrate. Mitochondria were from line 31 MS. Reaction mixture (3 ml, pH 7.2) contained 10 mM α-ketoglutarate, 0.3 M mannitol, 0.01 M KCl, 5 mM $MgCl_2$, 0.01 M KH_2PO_4, 0.1 M Tris-Cl and 2.25 mg bovine serum albumin. Additions of ADP were in amounts of 100 μ mol. Data are averages of 5 experiments ± S.E. of the mean (Chance and Williams 1956)

Studies on the mitochondrial complement were more promising in solving the problem of heterosis prognosis. According to the biochemical interpretation of heterosis the enzyme activity of hybrids—as seen in the foregoing— is higher than that of the parents. This increased activity is reflected in the respiratory activity of mitochondria as well (Table 92, Sarkissian 1972).

The data of the table show that the degree of manifestation of heterosis depends on the culture medium oxidated by the mitochondria.

The activity appeared in the hybrids even when the oxidative effect of the mitochondrium mixture obtained from different parents was studied (Table 93).

The efficiency of the mitochondrium mixture agreed with the values of the hybrid (the decrease was not significant). The correspondence is likewise good in the experiments carried out with maize (Fig. 46).

Table 93

Complementation by 1 : 1 mixtures of mitochondria of parents and comparisons with hybrid mitochondria (Sarkissian 1972)

Substrate*	ADP : O ratio		Per cent increase of F_1 over parents
	31 MS + 28	31 MS/28	
α-ketoglutarate	5.4	5.8	65
Malate	3.5	3.8	72
Pyruvate + malate	3.2	3.4	30
Succinate	3.2	3.4	112
NADH	2.0	2.4	50

* Substrate concentrations were as in Table 92. Pyruvate + malate were 2 mM pyruvate + "sparker" amount of malate.

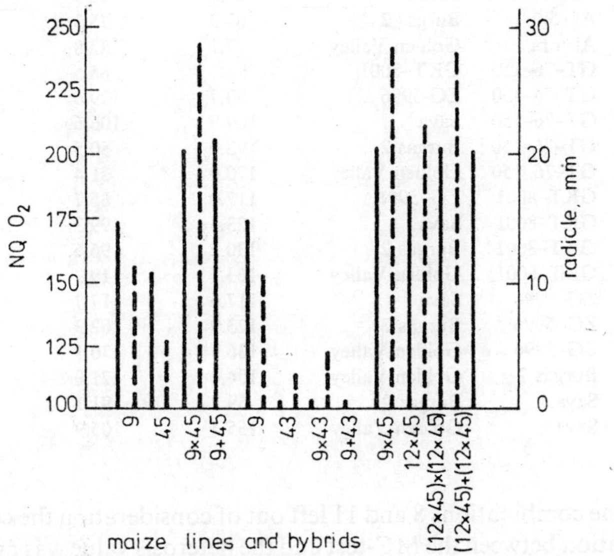

Fig. 46. Oxygen uptake by mitochondria of maize seedlings and length of radicle of the seedlings. Age of seedlings: 60 h; oxygen uptake: μl O_2/mg N/h (dark bar). Radicles lengths given in mm (striped bar). Maize lines shown by number only are WF9, Ohio 45, Ohio 43, West Virginia 12 (Sarkissian 1972)

The figure clearly shows that where no overdevelopment occurs in the growth of seedlings, there is no increase of activity even in the case of mitochondrial complementing.

Barabás et al. (1973) examined the combining ability of 21 wheat hybrids in thin stand (100 plant/m^2), in dense (stand/400 plant/m^2) and compared it with the results of the MC-test, as seen in Table 94.

Table 94

Miniplot grain yield and MC value (in Cytochrome-c-Oxydase units) of hybrids expressed as percent of better parent (Barabás et al. 1973)

Hybrid no.	Parent	Parent	Low seed rate	MC test	High seed rate
1	Aurora	GT-76-150	106.1	96.6	79.9
2	Aurora	GKT-8001	83.4	93.8	91.2
3	Aurora	ZG-5996	109.9	111.0	146.4
4	Aurora	Sava	93.1	73.9	122.6
5	Aurora	Burgas 2	63.2	75.3	125.5
6	Aurora	Golden Valley	77.1	82.8	117.1
7	GT-76-150	GKT-8001	91.1	65.5	125.7
8	GT-76-150	ZG-5996	90.7	129.6	78.2
9	GT-76-150	Sava	109.9	106.6	138.8
10	GT-76-150	Burgas 2	93.0	50.5	156.4
11	GT-76-150	Golden Valley	170.0	31.4	256.4
12	GKT-8001	ZG-5996	117.0	85.7	55.1
13	GKT-8001	Sava	123.5	99.3	183.4
14	GKT-8001	Burgas 2	130.2	96.6	126.7
15	GKT-8001	Golden Valley	163.1	119.2	153.1
16	ZG-5996	Sava	117.6	117.7	122.2
17	ZG-5996	Burgas 2	123.9	102.3	130.1
18	ZG-5996	Golden Valley	136.1	130.1	97.2
19	Burgas 2	Golden Valley	104.0	121.0	142.7
20	Sava	Burgas 2	69.2	91.1	114.6
21	Sava	Golden Valley	155.4	105.9	257.3

With the combinations 8 and 11 left out of consideration the coefficient of correlation between the MC-test and the heterosis value was $r = 0.61$ for the thin and $r = 0.1$ for the dense stand. The correlation is also weakened by the fact that some combinations show hybrid superiority in dense while others in thin stands.

Further, the method has been used to test a large number of maize combinations (Tsotsis 1972, personal communication). In maize breeding the method has not proved to be useful.

However, these partly negative experiences do not reduce the value of the method for, prognoses are not meant to give absolute results. A prognosis of the different gene effects accompanied by various methods of examination (analysis of B-vitamin and auxin contents in seed, MC-test, isotope studies etc.) might be most useful in the practice of breeding.

According to the opinion of Whaley (1952) the hybrid vigour is the result of an interaction between genes, metabolism and environment. The physiological genetic investigations have explored many factors of higher efficiency. So it is timely to apply this knowledge in producing hybrids with substantially higher yielding ability.

Conclusion

From the data compiled here breeders may realize how many biological variants are represented in the cultivars produced by them. Every effort should be made to utilize these hardly exploited possibilities in practice.

The plant grower becomes aware of the fact that he works with varieties and not with species. The variety is a sensitive biological system that can be successfully used only with a profound knowledge of its responses.

As pointed out in the introduction, often no data are available on the inheritance of the different processes, and often it is only on the variability of species and varieties that information can be given which is also essential. Vavilov's (1935) law concerning the homologous series suggests that the different methods of breeding should be used to come closer to the values of better varieties within a species and of species within genera possessing more efficient biological systems.

There is a great scope for us in this field to contribute to combatting world hunger and in raising the living standards of our nations.

Bibliography

Alvim, P. T. and Kozlowski, T. T. (1977): Ecophysiology of Tropical Crops. Academic Press, New York, San Francisco, London.

Andrejenko, S. and Kuperman, F. M. (1961): A kukorica élettana (Physiology of Maize). Mezőgazdasági Kiadó, Budapest.

Austin, R. B. and Jones, H. G. (1975): The Physiology of Wheat (Part III). Reprint. Plant Breeding Institute, Cambridge.

Badawi, A. M. (1973): Effect of some pesticides on agronomical characters and cytological abnormalities in *Pisum sativum* L. Gödöllő, Dissertation.

Bajai, I. and Barabás, Z. (1962): A takarmánycirkok nemesítése és termesztése (Breeding and cultivation of fodder sorghum). OMGK 1–120.

Baker, D. N., Hesketh, I. D. and Duncan, W. C. (1972): Stimulation of growth and yield in cotton. I. Gross photosynthesis, respiration and growth. *Crop Sci.* **12:** 431–435.

Baldy, Ch. M. (1974): Quelques reflexions concernant les caractères du rendement des blés. *Ann. Amélior. Plantes* **24:** 2, 193–199.

Bálint, A. (1967): Heterózis és mutáció a kukoricában (Heterosis and Mutation in Maize). Akadémiai Kiadó, Budapest.

Bálint, A. (1971): Növénynemesítés és kemizálás (Plant breeding and chemization). *Növénytermelés* **20:** 1, 85–92.

Bálint, A. (1976): A növénynemesítés alapjai (Fundamental Plant Breeding). Mezőgazdasági Kiadó, Budapest.

Bálint, A. (1977): Gazdasági növényeink produkciógenetikája (Production Genetics of Field Crops). Akadémiai Kiadó, Budapest.

Balla, L. and Szunics, L. (1978): Búzafajták szem : szalma arányának vizsgálata (Grain Straw Ratio in Wheat Varieties). *Növénytermelés* **27:** 6, 505–512.

Barabás, Z., Sági, F. and Kertész, Z. (1973): Special ways for hybrid wheat breeding. In: Fourth Int. Wheat Genetics Symposium, Columbia, Missouri, USA, 323–328.

(Baranovskii, A. G. and Savka, E. S.) Барановский, А. Г.–Савка, Е. С. (1974): Особенности продуктивности фотосинтеза у растений подсолнечника в онтогенезе. В: Обмен веществ и продуктивность растений; Изд. «Штиинца», Кишинёв.

(Bedrov, N. G.) Бедров, Н. Г. (1975): Селекция яровой пшеницы на устойчивость к весенней-летней засухе с учётом некоторых морфофизиологических показателей. В: Физиолого-генетические основы повышения продуктивности зерновых культур; Изд. «Колос», Москва, 291–299.

(Belozerskii, A. N. and Urison, S. O.) Белозерский, А. Н.–Урисон, С. О. (1959): Нуклеотидный состав дезокси-рибонукленновых и рубонукленновых кислот некоторых высших растений. АН СССР, 125, 5: 1144–47.

Bingham, I. (1972): Physiological objectives in breeding for grain yield in wheat. In: Proc. VI. Congress of EUCARPIA, Cambridge, 15–30.

Björkman, O., Troughton, I. and Nobs, M. (1974): Photosynthesis in relation to leaf structure. In: Basic Mechanism in Plant Morphogenesis. Brookhaven Symposia in Biology **25:** 206–226.

Boland, R. and Pickett, E. E. (1972): Breeding for Mg, Ca, K and P content in tall fescue. *Crop Sci.* **17:** 433–438.

Borlaug, N. E. (1972): Wheat. In: CIMMYT Ann. Report. Mexico, 84–91.

Borojevic, S. (1973): Canopy structure of different wheat genotypes in relation to the yield of grains. In: 4th International Wheat Genetics Symposium, Columbia, USA, 773–780.

(Botnar, L. S. and Soboleva, L. A.) Ботнар, Л. С.–Соболева, Л. А. (1974): Особенности фотосинтеза и дыхания межсортовых гибридов. В: Обмен веществ и продуктивность растений; Изд. «Штиинца», Кишинёв.

Böjtös, Z. (1976): Possibilities of the use of heterosis in lucerne hybrids made with a cytoplasmic male sterility system. In: Jánossy, A. and Lupton, F. C. H. (Ed.): Heterosis in Plant Breeding. Akadémiai Kiadó, Budapest, 213–218.

Brim, C. A. (1973): Quantitative genetics and breeding. In: Caldwell, B. A. (Ed.): Soybeans Improvement, Production and Uses. *Am. Soc. of Agron. Inc.* Madison, Wisconsin, USA, 155–186.

Bunting, E. S. (1973): Crop physiological studies in maize. In: VIIth Meeting of the Maize and Sorghum Section of EUCARPIA, 14.

Burt, R. L. (1965): The influence of reduced temperatures after emergence of the subsequent growth and development of the potato. *Eur. Potato* **1:** 8,104–114.

Burton, G. W., Prive, G. M. and Jackson, J. E. (1957): Studies of drought tolerance and water use of several southern grasses. *Agron. Jour.* **49:** 498–503.

Chandler, R. F. (jr.) (1969): Plant morphology and stand geometry in relation to nitrogen. In: P.C. Dinauer (Ed.): Physiological Aspects of Crop Yield. Publ. Amer. Soc. of Agronomy Madison, Wisconsin, USA, 236–285.

Chang, T. T. and Vergera, B. S. (1971): Ecological and genetic aspects of photoperiod-sensitivity and thermo-sensitivity in relation to the regional adaptability of rice varieties. *Int. Rice Comm. Newsl.* XX. 1–10.

Chatterton, N. I. (1976): Photosynthesis of 22 alfalfa populations differing in resistance to diseases, insect pests, and nematodes. *Crop Sci.* 833–834.

Clark, R. B. (1973): Efficiency of corn inbreds and sorghum lines for nutrient elements. In: Proc. of the 28th Annual Corn and Sorghum Res. Conf., Chicago, 144–160.

Crosbie, T. M., Mock, I. I. and Pearce, R. B. (1977): Variability and selection advance for photosynthesis in Iowa Stiff Stalk Synthetic maize population. *Crop Sci.* **17:** 511–514.

Cselőtei, L. (1964): A zöldségnövények vízhasznosítása (Water utilization of vegetables). Agrártud. Egyetem Mg. Tud. Karának Közl., Budapest, 203–226.

Cselőtei, L. (1978): Új irányok és feladatok a növények vízellátásában (New trends and tasks in the water supply of plants). *Agrártudományi Közlemények* **37:** 1, 45–57.

Damisch, W. (1973): Beiträge zur Ertragphysiologie des Getreides. 2. Mitt. Die Attraktions- und Produktionsbeziehungen in der Kornfüllungsperiode der verschiedenen Winterweizengenotypen. *Arch. Züchtungsforsch.* 3–4, 285–296.

Dark, S. O. S. (1967): A genetic effect on the progeny of tomato plants sprayed with 2,4-D. *J. Nat. Inst. Agr. Bot.* **11:** 1, 199–204.

Deckard, E. L., Lambert, R. J. and Hageman, R. H. (1973): Nitrate reductase activity in corn leaves as related to yields of grain and grain protein. *Crop Sci.* **13:** 343–350.

Deckard, E. L., Lucken, K. A., Joppa, L. R. and Hammond, J. J. (1977): Nitrate reductase activity, nitrogen distribution, grain yield and grain protein of tall and semidwarf nearisogenic lines of *Triticum aestivum* and *T. turgidum*. *Crop Sci.* **17:** 293–299.

DeLôughery, R. L. and Crookston, R. K. (1979): Harvest index of corn affected by population density, maturity rating and environment. *Agron. Jour.* **71:** 4, 577–580.

Demarly, Y. (1977): Génétique et l'amélioration des plantes. Masson, Paris, New York, Barcelona, Milan.

Derera, N. F. and Stoy, V. (1973): Varietal differences in photosynthesis efficiency of the awns. In: 4th International Wheat Genetics Symposium, Columbia, USA, 791–796.

Donald, C. M. (1968): The breeding of crop ideotypes. *Euphytica* **17:** 385–403.

(Dorokhov, B. L. and Baranina, I. I.) Дорохов, Б. Л.–Баранина, И. И. (1976): Фотосинтез озимой пшеницы при различном минеральном питании. Изд. «Штиинца», Кишинёв.

Duarte, R. and Adams, M. W. (1963): Component interaction in relation to expression of a complex trait in a field-bean cross. *Crop. Sci.* **3:** 185–186.

East, E. M. (1908): Inbreeding in corn. Report. Connecticut Agric. Expt. Stat for 1907. 419–428.

El-Sharkawy and Hesketh, I. (1965): Photosynthesis among species in relation to characteristics of leaf anatomy and CO_2 diffusion resistance. *Crop Sci.* **5:** 517–521.

Evans, L. T. (Ed.) (1975): Crop Physiology. Cambridge University Press.

(Fadeeva, T. S., Lutova, L. A. and Ageeva, L. A.) Фадеева, Т. С.–Лутова, Л. А.–Агеева, Л. А. (1975): Внутрупопиляционная изменчивость по изоферментному составу пероксидазы у сортов и инбредных линий редиса. *Генетика,* **11:** 11, 15–21.

Farkas, G. L. and Király, Z. (1958): Enzymological aspects of plant diseases *Phytopath. Z.* **31:** 251–272.

Fedak, G. and Rajhathy, T. (1972): Isosyme studies in hybrid barley. *Can. J. Plant Sci.* **52:** 751–756.

Fischer, H. E. (1978): Heterosis. VEB Gustav Fischer Verlag, Jena.

Frey, K. J. (1949): The inheritance of protein and certain of its components in maize. *Agr. J.* **41:** 113–117.

(Fyodorov, P. S.) Фёдоров, П. С. (1968): Биофизические и физиологические основы гетерозиса кукурузы. Изд. «Киргизистан» Фрунзе.

Gáspár, L. (1963): Contribution to the biology of hybrid vigour. In: Symposium on Maize Breeding and Production, Martonvásár, 59–66.

Gotlin, I. and Pucaric, A. (1973): LAI and grain yield of different inbred lines in relation to plant density. In: VIIth Meeting of Maize and Sorghum Section of EUCARPIA 5.

Grant, W. F. (1964): Cytogenetic effect of pesticides. Proceedings of the Tenth Annual Meeting of the Canadian Society of Agronomy, Frederiction, New Brunswick, June 22—25.

Grant, W. F. (1970): Pesticides and heredity. *Macdonald Journal* **31:** 211–214.

Grogan, C. O., Zuber, M. S., Brown, H. E., Whitehead, M. D. and Stanway, V. M. (1959): Effect of fungicides and insecticides on the germination of corn after storage. *Plant Disease Reporter* **43:** 10, 1132–1137.

Győrffy, B., I'só, I. and Bölöni, I. (1965): Kukoricatermesztés (Maize Growing). Mezőgazdasági Kiadó, Budapest.

Hageman, R. H., Lambert, R. I., Loussaert, D., Dalling, M. and Klepper, L. A. (1974): Nitrate and nitrate reductase as factors limiting protein synthesis. Urbana, USA Report 39.

Hajós-Novák, M. and Bálint, A. (1978): Sejtosztódási rendellenességek a nagy műtrágya-adagok hatására a kukoricánál (Disorders of cell division in response to high rate fertilization in maize). *Növénytermelés* **27:** 2, 97–102.

Hargitay, L. (1976): Különböző növényvédőszerek egyedi és szinergista hatásának genetikai és élettani vizsgálata kukoricán (Genetic and Physiological Investigation of Individual and Synergic Effects of Different Pesticides on Maize). Gödöllő, Dissertation.

Hargitay, L., Kovács, G. and Sebestyén, E. (1978): A Rawixen 50WP herbicid okozta sejtosztódási rendellenességek a kukoricában (*Zea mays* L.) (Cell division disorders induced by the herbicide Rawixen 50WP in maize (*Zea mays* L.). *Növénytermelés* **27:** 3, 193–199.

Harper, I. L. (1977): Population Biology of Plants. Academic Press, London, New York, San Francisco.

Harvey, P. H. (1939): Hereditary variation in plant nutrition. *Genetics* **24:** 437–461.

Hayashi, K. (1978): Perspective on agricultural technology for the twenty-first century. *Technocrat* **11:** 1, 12–30.

Hayashi, K. and Ito, H. (1962): Studies on the form of plant in rice varieties with particular reference to the efficiency in utilizing sunlight. *Proc. Crop Sci. Soc. Japan* **30:** 329–333.

Haynes, P. H., Spence, I. A. and Walter, C. I. (1967): The use of physiological studies in the agronomy of root crops. Proc. Int. Symp. Trop. Root Crops, 1st. 1970. I, 1–17.

Heichel, G. H. and Musgrave, R. B. (1969): Varietal differences in net photosynthesis of *Zea mays* L. *Crop Sci* **9:** 485–486.

Heidrich-Sobrinho, E. and Cordeiro, A. R. (1975): Codominant *isoenzymic alleles* as markers of genetic diversity correlated with heterosis in maize *(Zea mays.) Theoretical and Applied Genetics* **46:** 4, 197–199.

Hess, D. (1968): Biochemische Genetik. Springer Verlag, Berlin, Heidelberg, New York.

Hess, D. (1974): A gének menetrendje (The Time-table of Genes). Natura, Budapest.

Hess, D. (1979): Növényélettan (Plant Physiology). Natura, Budapest.

Hoffer, G. N. (1926): Some differences in the functioning of selfed lines of corn under varying nutritional conditions. *Jour. Amer. Soc.Agron.* **18:** 322–334.

Hoffmann, P. (1975): Photosynthese. Akademie-Verlag, Berlin.

Hooker, A. L. (1967): *Helminthosporium maydis,* a Challenge for the Maize Breeder. In: Proc. of the 6th Meeting of the Maize and Sorghum Section of EUCARPIA. Freising, 333–367.

Horváth, P. (1963): A szudánifű. In: Nemesített növényfajtákkal végzett országos fajtakísérletek eredményei (The sudan grass. In: Results of National Variety Trials with Improved Plant Varieties). Mezőgazdasági Kiadó, Budapest, 285–292.

Humphries, E. C. (1969): The dependence of photosynthesis on carbohydrate sinks: current concepts. Proc. Int. Symp. Tropical Root Crops, Univ. of West Indies **1:** II, 34–35

Ishizuka, Y. (1969): Engineering for higher yields. In: Dinauer, R. C. (Ed.) Physiological aspects of crop yield. Publ. Amer. Soc. of Agronomy Madison, Wisconsin, USA, 16–26.

I'só, I. (1969): Tenyészterület-kísérletek különböző hibridekkel. In: Kukoricatermesztési kísérletek 1965–1968 (Spacing experiments with various hybrids. In: Maize Growing Experiments 1965–1968). Akadémiai Kiadó, Budapest.

Ivins, I. D. and Bremner, P. M. (1964): Growth, development and yield in the potato. *Outl. Agric.* **4:** 211–217.

Johnson, D. R. and Major, D. I. (1979): Harvest index of soybeans as affected by planting date and maturity rating. *Agronomy Journal* **71:** 4, 538–541.

Jugenheimer, R. W. (1958): Hybrid Maize Breeding and Seed Production. FAO, Rome.

Kádár, A. and László, I. (1975): A kukoricafajták herbicid érzékenységének vizsgálata (Herbicide sensitivity of maize varieties). *Magyar Mezőgazdaság* **30:** 44, 10–11.

Kiesselbach, T. A. (1926): The comparative water economy of selfed lines of corn and their hybrids. *Jour. Amer. Soc. Agron.* 335–344.

Koltay, Á. and Balla, L. (1975): Búzatermesztés és nemesítés (Wheat Growing and Breeding). Mezőgazdasági Kiadó, Budapest.

(Konarev, V. G.) Конарев, В. Г. (ред.) (1971): Вопросы биохимии гетерозиса у растений. АН СССР, «Башкирский филиал», Уфа.

Koo, F. K. S. (1972): Mutation breeding in soybeans. In: Induced Mutation and Plant Improvement. IAEA, Vienna, 285–292.

Kostoff, D. (1939): Effect of the fungicide "Granosan" on atypical growth and chromosome doubling in plants. *Nature* **144:** 334.

Kovács-Schneidler, M. (1972): Study on the physiological and biochemical bases of combining ability in maize lines and hybrids. *Acta Agronomica* **21:** 1–2, 89–100.

Kraljevic-Balalic, M. (1973): Inheritance of leaf area in vulgare wheat. In: 4th International Wheat Genetics Symposium, Columbia, USA, 773–780.

(Kuperman, F. M.) Куперман, Ф. М. (1977): Морфофизиология растений. «Высшая школа», Москва.

Kurnik, E. (1970): Étkezési és takarmányhüvelyesek termesztése (Production of Consumption and Fodder Legumes). Akadémiai Kiadó, Budapest.

Kurnik, E. (1976): A növénynemesítés távlati célkitűzései (Long-term objectives of plant breeding). *Agrártud. Közlemények* **35:** 1–3, 231–237.

Kurnik, E. (1978): Energia-szempontok a növénynemesítésben (Energy questions in plant breeding). *Agrártud. Közlemények* **37:** 1, 59–70.

Li, S. L. and Rédei, G. P. (1969): Direct evidence for models of heterosis provided by mutants of Arabidopsis blocked in the thiamine pathway. *Theor. and Appl. Genetics* **39:** 68–72.

(Lobasheva, M. E. and Inge-Vechtomova, S. G.) Лобашева, М. Е.–Инге-Вечтомова, С. Г. (1976): Физиологическая генетика. «Медицина», Ленинград.

Loomis, R. S. and Williams, W. A. (1972): Plant conformation and yield. In: Induced Mutation and Plant Improvement. IAEA, Vienna, 13–25.

Loomis, R. S., Williams, W. A. and Hall, A. E. (1971): Agricultural productivity. *Plant Physiology* **22:** 431–468.

Lorenz, H. (1975): Relationships between yield and free amino acids in hybrid lines of corn. *Zeit. Pflanzenzüchtung* **75:** 1, 10–23.

Lupton, F. G. H. (1966): Translocation of photosynthetic assimilates in wheat. *Ann. Appl. Biol.* **57:** 335–364.

Lupton, F. G. H. (1972): Further experiments on photosynthesis and translocation in wheat. *Ann. Appl. Biol.* **71:** 69–79.

Lupton, F. G. H. (1976): The physiological basis of heterosis in Wheat. Heterosis in Plant Breeding. EUCARPIA, Budapest, 1974. Akadémiai Kiadó, Budapest, 71–79.

Lupton, F. G. H., Ali, A. M. M. and Subramanion, S. (1967): Varietal differences in growth parameters of wheat and their importance in determining yield. *J. Agric. Sci. Camb.* **69:** 111–123.

Ludlow, M. M. and Wilson, G. L. (1971): Photosynthesis of tropical pasture plants. *Aust. J. Biol. Sci.* **24:** 449–470.

Lynese, A. S. (1936): Varietal differences in the phosphorus feeding capacity of plants. *Plant Physiol.* **11:** 665–688.

Manzyuk, S. G. (1973): The role of physiologically active substances in the expression of heterosis in maize (Zea mays L.). In: Proc. of the 7th meeting of the Maize and Sorghum of EUCARPIA, Zagreb, Part 2. 1–9.

Marsálek, L. (1972): Unterschiede in der Aktivität einziger Enzyme bei den Genotypen einer Doppelhybriden von Mais. *Zeit. Pflanzenzüchtung* **67:** 4, 305–312.

Matzkov, F. F. and Manzyuk, S. G. (1962): Fitohormon és vitamin típusú anyagok szerepe a kukorica heterózis jelenségeiben (Role of phytohormone- and vitamin type substances in the heterosis phenomena of maize). *MTA Biol. Csop. Közl.*, **5:** 1–2, 63–76.

Maximov, N. A. (1929): Experimentelle Änderungen der Länge der Vegetationsperiode bei der Pflanzen. *Biol. Zb.* **49:** 513–543.

(Maximov, N. A. and Kohanovskaya, L.) Максимов, Н. А.–Кохановская, Л. (1921): Experiments on transpiration of plants in subalpine zone. Publ. of the First Congress of Russian Botanists in Petrograd. (In Russian)

Miller, I. H. (1958): Differential response of several inbreds and single crosses of corn *(Zea mays)*, to 2,4 dichlorophenoxyacetic acid. *Diss. Abst.* **18:** 4, 1197–1198.

Murata, Y. (1961): Studies on the photosynthesis of rice plants and its culture significance. *Bull. Nat. Inst. Agr. Sci. Japan* Ser. D. **9:** 1–169.

Murata, Y. (1969): Physiological responses to nitrogen in plants. In: Dinauer R. C. (Ed.) Physiological Aspects of Crop Yield. Publ. Amer. Soc. of Agronomy Madison, Wisconsin, USA, 235–259.

Murata, Y. and Matsushima, S. (1975): Rice. In: Evans L. T. (Ed.) Crop physiology. Cambridge University Press.

Murata, Y. and Togari, Y. (1972): Analysis of the effect of climatic factors upon the productivity of rice at different localities in Japan. *Proc. Crop Sci. Soc. Japan* **41:** 372–387.

Musgrave, R. B. (1971): Photosynthetic efficiency in corn. In: Proceedings of the 26th Annual Corn and Sorghum Research Conference, Chicago, 186–191.

Nagy, B. (1979): Az agrárkemizálás környezetvédelmi helyzete és problémái (Environment protection aspects of agricultural chemization). *Növényvédelem* **15:** 2, 49–59.

Nelson, W. L. (1966): Fertilizer practices for tomorrow's agriculture. In: Proceedings of the 19th Annual Hybrid Corn Industry-Research Conference, Chicago, 14–22.

(Nichiporovich, A. A.) Ничипорович, А. А. (1957): О некоторых физиологических основах системы дифференцированной агротехники. Изд. АН СССР, Москва.

Nichiporovich, A. A. (1967): A fotoszintézis és a bőséges termés (Photosynthesis and Abundant Yield). Mezőgazdasági Kiadó, Budapest.

Nickell, L. B. (1977): Sugarcane. In: Alvim, P. T. and Kozlowski, T. T. (Eds): Ecophysiology of Tropical Crops. Academic Press, New York, San Francisco, London, 89–111.

Nygren, A. (1949): Cytological studies of the effects of 2,4–5-T on Allium cepa. *Ann. Royal Agr. Coll.* Sweden **16:** 723–728.

Osada, A. (1966): Relationship between photosynthetic activity and dry matter production in rice varieties, especially as influenced by nitrogen supply. *Bull. Nat. Inst. Agr. Sci. Japan*, Ser. D, **14:** 117–188

Pap, E. (1953): Heterózisos kukoricanemesités (Hybrid corn breeding). *MTA Agrártud. Oszt. Közleményei* **2:** 1, 1–14.

Pap, L., Kovács, G. and Bálint, A. (1977): Az Igrán 80 WP hatása a kukoricalevelek in vivo nitrátreduktáz (NR) aktivitására (Effect of Igrán 80 WP on in vivo nitrate reductase (NR) activity in maize leaves). *Növénytermelés* **26:** 6, 481–490.

Pál, I. (1973): Általános növényélettan (General plant physiology). Agrártud. Egyetem Mg. Tud. Kar, Gödöllő, lecture notes.

Peet, M. M., Bravo, A., Wallace, D. H. and Ozbun, I. L. (1977): Photosynthesis, stomatal resistance, and enzyme activities in relation to yield of field-grown dry bean varieties. *Crop Sci.* **17**: 287–293.

Pendleton, I. W. (1968): Light relationship and corn plant geometry. In: Proc. of the 23th Annual Corn and Sorghum Research Conf., ASTA, Chicago 91–96.

Petinov, N. Sz. (1965): Az öntözött mezőgazdasági növények fiziológiája (Physiology of Irrigated Farm Crops). Mezőgazdasági Kiadó, Budapest.

Petrasovits, I. and Balogh, J. (1975): Növénytermesztés és vízgazdálkodás (Plant Industry and Water Economy). Mezőgazdasági Kiadó, Budapest.

Piovaric, A. and Vidovic, I. (1973): Leaf angle in corn (*Zea mays* L.) breeding. In: Proc. of the 7th Meeting of Maize and Sorghum Section EUCARPIA-Zagreb. 11.

Planchon, C. (1974): Assimilation nette par unité de surface chez diverses éspéces du genre Triticum. *Ann. Amélior. Plantes* **24**: 2, 201–207.

Pollhamer, Zs. (1973): A búza minősége különböző agrotechnikai kísérletekben (The Quality of Wheat in Various Agrotechnical Experiments). Akadémiai Kiadó, Budapest.

Pollmer, W. G., Eberhardt, D., Klein, D. and Dhillon, B. S. (1979): Genetic control of nitrogen uptake and translocation in maize. *Crop Sci.* **19**: 1, 81–86.

Powell, R. D. (1968): High rates of fertilization for corn. In: Proc. of the 23th Annual Corn and Sorghum Res. Conf., Chicago, 27–32.

Rabideau, G. S., Whaley, W. G. and Heimsch, Ch. (1950): The absorption and distribution of radioactive phosphorus in two maize inbreds and their hybrid. *Am. Journ. of Bot.* **37**: 93–99.

Rabson, R., Bhatia, C. R., Mitra, R. K. (1978): Crop productivity, grain protein and energy. In: Seed Protein Improvement by Nuclear Techniques. IAEA, Vienna, 3–20.

Rajki, E. and Rajki, S. (1969): A növekedési habitus monoszómás analízise az ősziesítés folyamán (Monosome analysis of growth habit in the course of autumnization). *Növénytermelés* **18**: 1, 89–91.

Rao, K. P., Rains, D. W., Qualset, C. O. and Huffaker, R. C. (1977): Nitrogen nutrition and grain protein in two spring wheat genotypes differing in nitrate reductase activity. *Crop Sci.* **17**: 2, 283–286.

Robbins, W. J. (1941): Factor Z in hybrid maize. *Bull. Torray Bot. Club* **68**: 222–228.

Romanovskaya, O., Gubar, G., Irbe, I. and Straume, O. (1968): Change of intensity of photosynthesis and respiration in plants due to the selective herbicidic action of phenazone. *Biologia* **8** (253), 90–97.

Rosielle, A. A. and Frey, K. I. (1977): Inheritance of harvest index and related traits in oats. *Crop Sci.* **17**: 1, 23–28.

Russel, W. A. and Balko, L. G. (1980): Response of corn inbred lines and single crosses to nitrogen fertilizer. In: Thirty-Fifth Annual Corn and Sorghum Research Conference. ASTA Chicago, 48–67.

Sammons, D. I., Peters, D. B. and Hymowith, T. (1978): Screening soybeans for drought resistance. I. Growth Chamber Procedure. *Crop Sci.* **18**: 6, 1050–1055.

(Samokhvalov, G. K.) Самохвалов, Г. К. (1972): Физиология питания и развития растений. Изд. «Харьковского Университета», Харьков.

Sarkissian, I. V. (1972): Mitochondrial polymorphism and heterosis. *Z. Pflanzenzüchtung* **67**: 1, 53–64.

Sayre, I. D. (1955): Mineral nutrition of corn. In: Sprague, G. F. (Ed.): Corn and Corn Improvement Academic Press, New York, 293–314.

Schwartz, D. (1962): Genetic studies on mutant enzymes in maize. II. On the mode of synthesis of the hybrid enzymes. *Proc. of the Nat. Acad. of Science* **48:** 750–756.

Serbanescu, E. (1966): Cercetari fiziologice la plante hibride si la formele lor parantele. *Studii si corcetari de biologie* **18:** 5, 461–469.

(Semenenko, G. I.) Семененко, Г. И. (1964): К биохимии обмена нуклеиновых кислот у высших растений. Изд. «Харьковского Университета», Харьков.

Shannon, I. C. (1978): Physiological factors affecting starch accumulation in corn kernel. 33. Annual Corn and Sorghum Conference, Chicago, 78–94.

Singh, H. N., Singh, H. R. and Vaishampayan, A. (1979): Toxic and mutagenic action of the herbicide alachlor (Lasso) on various strains of the nitrogen-fixing blue green alga Nostoc muscorum and characterization of the herbicide-induced mutants resistant to methylamine and L-methionine-DL-sulfoximine. *Environmental and Experimental Botany* **19:** 1, 5–12.

Singh, I. I. and Singh, P. P. (1970): Response of some sugarcane varieties to N-application in North-India. *The International Sugar Journal* June.

(Sysoev, A. F. and Semenyuk, B. F.) Сысоев, А. Ф.–Семенюк, Б. Ф. (1975): Биохимические показатели потенциальной продуктивности сортов мягкой озимой пшеницы. В. кн: Физиолого-генетические основы продуктивности зерновых культур. «Колос», Москва, 94–102.

Sleper, D. A., Garner, G. B., Asay, K. H., Boland, R. and Pickett, E. E. (1977): Breeding for Mg, Ca, K and P content in tall fescue. *Crop Sci.* **17:** 3, 433–438.

Smith, S. N. (1934): Response of inbred lines and crosses in maize to variations of nitrogen and phosphorus supplied as nutrients. *Jour. Amer. Soc. Agron.* **26:** 785–804.

Somos, A. (1966): A paprika (Paprika). Akadémiai Kiadó, Budapest.

Stoy, V. (1969): Interrelationship among photosynthesis, respiration and movement of carbon in developing crops. In: Dinauer, R. C. (Ed.): Physiological Aspects of Crop Yield. Amer. Soc. of Agronomy Madison, Wisconsin USA, 185–206.

Stoy, V. (1976): Source and sink properties as related to yield in different barley genotypes. In: Barley Genetics III. Symposium, Carching, 1975, 641–648.

Strebeyko, P. (1973): Influence of water deficit on photosynthetic activity of some crop plants. *Hodowla roslin, aklimatyzacija i nasiennictwo* **17:** 5, 413–416.

Stubbe, K. and Pirschle, K. (1940): Über einen monogen bedingten Fall von Heterosis bei Antirrbhinum majus. *Ber. Deutsch. Bot. Gesl.* **58:** 546–558.

Suresh K. Sinha and Renu Khanna (1975): Physiological, biochemical and genetic basis of heterosis. *Advances in Agronomy,* **27:** 123–174.

Sváb, J. (1971): A populációgenetika alapjai (Fundamentals of Population Genetics). Mezőgazdasági Kiadó, Budapest.

Swaminathan, M. S. (1972): The role of nuclear techniques in agricultural research in developing countries. In: Peaceful Uses of Atomic Energy. Vol. 12, 3–32.

Swaminathan, M. S. (1973): The plant population explosion. *Ceres* **6:** 2, 11–15.

Szalai, I. (1974): Növényélettan I–II (Plant physiology). Tankönyvkiadó, Budapest.

Tamássy, I. (1973): A bioregulációs kutatások jelentősége a kertészeti növénynemesítésben és az alkalmazott kertészeti tudományok területén (Importance of bioregulation studies in plant breeding and applied horticultural sciences). *Agrártud. Közl.* **32:** 3–4, 373–390.

Tarchevskii, I. A., Chicov, V. I., Suleomanova, A. Yu., Ivanova, A. P. and Andrianova, Y. E. (1975): Photosynthesis, assimilation numbers and photosynthate translocation in wheats of different stalk length. In: Nasyrov, Y. S., Sesták, Z. and Iunt, W. (Eds) In: Genetic Aspects of Photosynthesis. The Hague, 363–366.

(Tarchevskii, I. A., Chikov, V. I. and Ivanova, A. P.) Тарчевский, И. А.–Чиков, В. И., Иванова, А.П. (1977): Особенности фотосинтеза и оттока ассимилятов у различных сортов яровой пшеницы. Генетика фотосинтеза. Изд. «Доныш», Душанбе 234–237.

Thomas, W. (1963): Chemical element accumulation in corn. In: Eighteenth Hybrid Corn Industry-Research Conference. American Seed Trade Association, Washington, 53–61.

(Tokarev, B. I. and Shumnyi, V. K.) Токарев, Б. И.–Шумный, В. К. (1977): Влияние у ячменья мутантов с пониженной нитратредуктазной активностью после обработки зерна этилметансульфонатом. *Генетика* **13,** 12, 2097–2103.

Tomkins, D. J. and Grant, W. F. (1972): Comparative cytological effects of the pesticides menazon, metrobromuron and tetrachloroisophatalonitrile in Hordeum and Tradescantia. *Can. J. Genet. Cytol.* **14:** 245–246.

Triplett, G. B., Van Keuren, R. W. and Walker, L. D. (1977): Influence of 2–4 D, Pronamide, and Simazine on dry matter production and botanical composition of an Alfalfagrass Sward. *Crop Sci.* **17:** 1, 61–65.

Tsotsis, B. (1972): Objectives of industry breeders to make efficient and significant advances in the future. Proc. of the 27th Annual Corn and Sorghum Res. Conf. Chicago, 93–107.

Ubrizsy, G. (1972): A mezőgazdasági munka kemizálása és a környezet. (Agricultural chemization and environment.) *Agrártud. Közlemények* **31:** 3–4, 431–449.

Újvárosi, M. (1969): Gyomnövényvizsgálatok Hungazinnal gyomirtott kukoricavetésekben (Weed studies in maize stands treated with the herbicide Hungazin). *Agrártud. Közlemények* **28:** 1–2, 1—18.

(Vavilov, N. I.) Вавилов, Н. И. (1935): Закон гомологических рядов в наследственной изменчивости. 11. издание «Сельхозгиз», Москва—Ленинград.

Vincent, A. (1970): Régulation du métabolisme par le système Phytochrome-Ascorbique. EUCARPIA Proc. Dijon oct. 1970. 11–80. ed INRA B. V. 1540. Dijon, France.

Voldeng, H. D. (1971): Growth in *Zea mays*. D. Phil. Thesis, Univ. of Oxford, England.

Vose, P. B. (1963): Varietal differences in plant nutrition. *Herbage Abstract* **33:** 1, 1—13.

Wallace, D. H., Ozbum, I. L. and Crookston, R. K. (1973): Physiological genetics and application in plant breeding. Cornell University, Ithaca, N. Y. (Manuscript).

Wallace, D. H., Ozbum, I. L. and Munger, H. M. (1972): Physiological genetics of crop yield. *Advances in Agronomy* **24:** 97–146.

Warner, R. L. and Kleinhofs, A. (1974): Relationships between nitrate reductase, nitrite reductase, and ribulose diphosphate carboxylase activities in chlorophyll-deficient mutants of barley. *Crop Sci* **14:** 5, 654–658.

Whaley, W. G. (1952): Physiology of gene action in hybrids. In: Gowen, I. W.: Heterosis, Ames, Iowa, 98–113.

Wittwer, S. H. (1974): Maximum production capacity of food crops. *Bio Science* **24:** 4, 216–224.

Wuu, K. D. and Grant, W. F. (1966): Morphological and somatic chromosomal aberrations induced by pesticides in barley *(Hordeum vulgare)*. *Can. J. Genet. Cytol.* **8:** 481–501.

Zelitch, I. (1971): Photosynthesis, photorespiration and Plant Productivity. Academic Press, New York.

Zwicker, R. (1976): Probleme und Stand der Arbeiten zu einer Frühdiagnose bei Winterweizen. Beiträge zur Züchtungforschung Getreide. Tag.-Ber., *Landwirtsch. Wiss.* DDR, Berlin **143:** 323–330.

Yamada, E. E. (1959): The nature of fertilizer response in japonica and indica rice varieties. *Int. Rice. Comm. Newslett.*, **8**: 14–19.

(Yakovlev, A. and Raskova, N.) Яковлев, А.–Раскова, Н. (1975): Активность сукциндехидрогеназ в самоопыленных линиях и гибридах кукурузы. *Вестник с-х наук*. **10**: 238: 68–71.

Yamaguchi, H., Watanabe, M., Sano, S. and Kanbayashi, Y. (1976): Studies on a partially asynaptic and upright leaf-areal mutant. In: *Gamma Field* Symposia No. 15. Japan 1–15.

Yoshida, S., Cock, I. H. and Parao, F. T. (1973): Physiological Aspects of High Yields. In: Rice Breeding IRRI, Los Banos, P. 455–468.

IAEA (1975): Tracer technique for plant breeding Vienna.

Subject index